THE

BOOK
OF
ELECTRONICS

THE

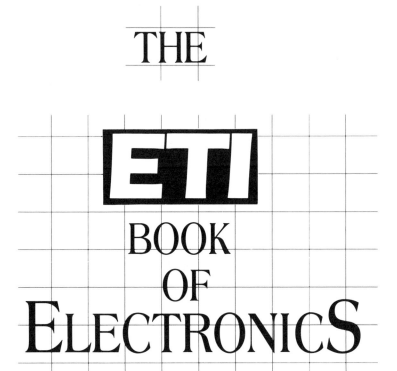

ETI
BOOK
OF
ELECTRONICS

DAVE BRADSHAW

ARGUS BOOKS

Argus Books
Argus House
Boundary Way
Hemel Hempstead
Herts. HP2 7ST
England

First published by Argus Books 1992

ISBN 0 85242 928 2

Phototypesetting by GCS, Leighton Buzzard. Printed and bound in Great Britain by Biddles Ltd, Guildford and King's Lynn

Dedication

Firstly, this book must be dedicated to my mother, without whom I would not have happened, never mind the book.

Secondly, it must be dedicated to colleagues — sadly all erstwhile colleagues: Pete, Phil, Ian, Helen, Steve, ... and even Rory, Gary and Paul. You taught me more than I'm prepared to admit!

And lastly, this book is dedicated to Katherine, long may you continue to teach me about myself.

Contents

CHAPTER 1

Introduction: Getting Started

The object of this book is to introduce you to the fascinating world of electronics. To prove that you can start using electronics with relatively little knowledge, I will begin by showing you how to make interesting and useful projects after the first few chapters.

Electronics works entirely by manipulating electricity, so Chapter 2 looks at the fundamentals of electricity itself. Chapter 3 introduces alternating electricity, and some special components that take advantage of its properties. Modern electronics has been made possible by semi-conductors and Chapter 4 begins describing what they are. Chapter 5 introduces transistors – the fundamental building blocks made with semiconductors, and Chapter 6 explains how large numbers – perhaps millions – of transistors are combined into microchips. Chapter 7 looks at transducers, devices through which all electronic circuits interact with the outside world. Chapter 8 discusses digital electronics which have made possible the building of sophisticated computers. Finally, Chapter 9 looks at how you can take your interest in electronics further.

For the remainder of this chapter, we will look at what *signals* are and why they are important, at *circuit diagrams* and how they are used, and then briefly we will talk about the sorts of *tools* and other equipment needed for electronics.

In this book, the maths has been kept to a simple minimum. In electronics, maths is both a tool and a shorthand used for representing real voltages and currents, so it is important to understand and be able to use the few basic formulae. They are mostly straightforward and become second nature after a bit of practice.

Signals

A signal is a piece of information in electrical form. Three examples of signals are the output from the cartridge of a record deck, the radio waves which bring us music and news with our breakfast, and the data output from a computer to a printer. Electronic circuits are used to alter these signals in some controlled way, making them more useful. Let us look at those three examples to see what the circuits have to do.

The signal from the record deck is very weak so the amplifier has to make it strong enough to drive the loudspeakers. Also, only those parts of the signal which correspond to audible sound have to be selected, and any which are below or above the range of human hearing must be cut out. And a correction must be applied to the remaining signal because, due to the recording system used, higher frequencies are recorded at higher volume levels than they should be played.

A radio signal consists of high frequency changes of voltage which are themselves further varied (or *modulated*) in either size or frequency by a much lower frequency signal. This lower frequency signal might be from a microphone or a record player at the radio station. First, the radio has to select the desired station's signal from hundreds of others (tune across the medium or short wave band at night to see how many there are); much of a radio's circuitry is devoted to accurately doing this. After this, the lower frequency signal is recovered and amplified so that it can drive the loudspeaker.

Computer signals consist of distinct 'ons' and 'offs'; 'on', usually referred to as high or 1, will be a fixed voltage, often 5 volts, or close, and 'off', usually referred to as low or 0, is usually zero volts, or close. A useful signal consists of groups of ons and offs forming a code that computers and printers can understand and deal with. At the printer, electronic circuits have to interpret these groups and then control the printer's mechanism to print patterns on the paper. These patterns might be letters or special graphic symbols.

Circuit diagrams

A circuit diagram (also called a *schematic*) shows how the different parts in an electronic circuit are connected together. There are conventions decreeing what symbols are used for the different components. Although there are at least two competing sets of symbols, they are not too dissimilar. There are other conventions which make circuit diagrams easier to follow. We usually start with the signal entering the circuit from the left-hand side and moving towards the right as it passes through different sections to the output on the right.

Nearly all circuits require some form of power – from a battery or from the mains via a power supply. The power is usually shown entering the circuit from the right. In simpler circuits we usually draw the power supply paths as continuous lines running from left to right. However, to make the complex circuits easier to follow, we often use symbols instead to indicate connection to the supply lines.

In all circuits, one symbol used a lot is the symbol for *common* (Fig. 1.1), the point in a circuit which is a reference point for all the signals. Using this

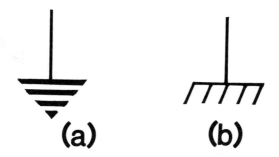

Fig. 1.1 The symbols for (a) ground or earth, and (b) common.

symbol removes the need to draw long lines just to connect up all the common points on circuit diagrams, making it much easier to see what is going on. Common should not be confused with *earth*, also called *ground* (*GND*), which literally means a point connected to the earth, often via the mains earth connection. In electronic circuits which use earth, it is usually also the common, at least for signals. However, many circuits do not connect to earth, and the common connection (as well as the rest of the circuit) is left 'floating', it can be at any voltage with respect to earth.

Some circuits have several different supplies, with different voltages and currents, and these will all share a single power common. There may be two commons, one for signals and one for the power supplies. Although the two are normally connected, this is just at one point. The reason for this is to keep the relatively strong supply voltages and currents as separate as possible from the much weaker signals, so that unwanted 'noise' signals in the supplies do not turn up in the signal.

Tools for electronics

You will find many of the tools you need in the average household toolkit. It will, however, prove useful to acquire smaller versions of long-nosed pliers, wire cutters and screwdrivers (all with insulated handles). Two items will have to be specially bought: a soldering iron and a multimeter.

Soldering iron
A typical soldering iron used for electronics is shown in Fig. 1.2. The 'size' of a soldering iron is its power, i.e. how much heat it can deliver to the bit. A good power rating for electronics is 15 to 18W. The iron should have interchangeable bits – I suggest getting a standard bit of 3 to 5mm diameter and a medium fine bit of 2 to 3mm.

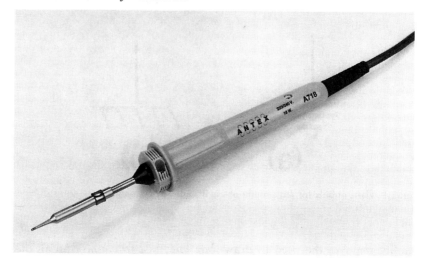

Fig. 1.2 A typical soldering iron (photograph courtesy Antex Electronics and Maplin Electronics).

Choose a soldering iron which uses the mains for its supply, not one which needs a special power supply or runs from a battery. Avoid soldering guns, as they are too heavy for fine work. Some irons have ceramic-lined shafts, so that the bit is insulated from earth; I have not found this necessary. There are temperature-controlled irons, using a high-power (say 50 watts) element and a thermostat – these are useful for experts but unnecessary for beginners.

There are two essential accessories: a stand for the iron that keeps the bit and element covered when the iron is hot but not being used, and a mains plug with a neon light. Without these, you will, sooner or later, burn yourself. Although soldering iron burns are rarely serious, they can be very painful and are worth avoiding. Soldering irons are not an instant fire hazard, but treat them with respect. Don't leave them around inflammable materials, and switch them off when not in use. They can singe many materials, including wood.

Multimeter
A multimeter is a general-purpose electrical measuring instrument which will measure steady and alternating voltages, steady and possibly alternating currents, and resistance.

There are two types of multimeter: analogue and digital. Analogue meters have a scale and pointer for readings, digital meters have an electronic numerical display. In general, you can buy a good analogue meter for less than the price of a poor digital one, so buy an analogue meter to start with. Later, you can buy a digital meter as both types have their uses.

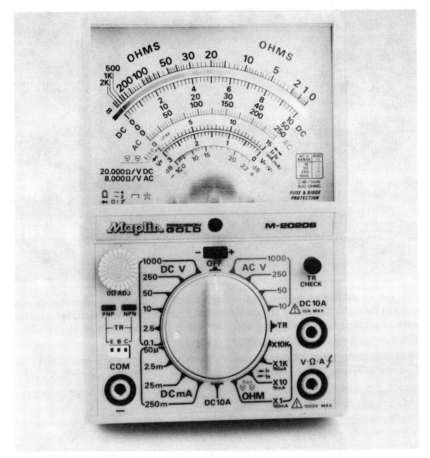

Fig. 1.3 A multimeter, showing the main features (photograph courtesy Maplin Electronics).

Many of the following terms may be unfamiliar; they are used here to help you select the right sort of meter, but are explained fully in later chapters.

The first criteria is the scale. If it seems clear and well laid out, this is the best. Some manufacturers put far too many scales onto their meters, with the result that you can never find the correct one to take the reading from. Choose a meter where you can see what is going on. The second or third DC volts range is the one I use the most. You will need one with a maximum reading, which is between 10 and 25 volts, and my ideal (rarely found) would be 15 volts. Also, a 5V DC range can be very useful for digital electronics. On the AC ranges, it is useful to have ranges of 10, 25 or 50 volts and a 250 volt.

The sensitivity of the meter tells us how much the meter affects the circuit

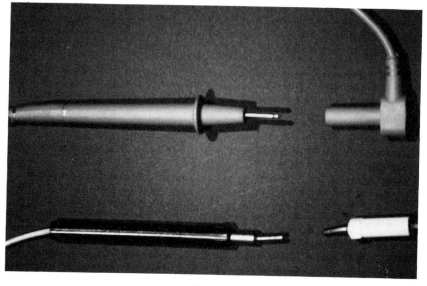

Fig. 1.4 Two different multimeter test prods (to the left), with the plugs that go with them to the right. The one at the top is fine, but the lead at the bottom makes it far too easy to touch live metal at either end, and is only fit for throwing away.

it is measuring, sometimes causing errors. The higher the sensitivity, the lower the errors, so get a meter with a sensitivity of 20,000 ohms per volt (or more, but this is not usually necessary) on its DC ranges. The sensitivity of AC ranges is not usually critical.

The resistance ranges should offer a reasonable selection with individual ranges having 100 to 200 ohms (or Ω), 1 to 2 kohms and 10 to 20 kohms at the centre of the scale. (Resistance ranges usually have zero at the right and infinite resistance at the left, so reading the centre of the scale is the best way of comparing them.) There should be at least one very low current range, say 50 or 100 microamps (uA or μA), with other ranges stretching from a few milliamps up to 1 or perhaps 10 amps (A). An AC current range of say 100mA, 500mA or 1A is relatively rare, but can be useful very occasionally.

The full scale deflection (FSD) of a meter means the maximum reading on a range. The jargon '10 volts FSD' means '10 volts maximum reading'. Meter accuracies are normally quoted as a percentage of the FSD, and you should look for at most 5%, and better (smaller) if you can afford it.

Finally, the meter's test leads are worth looking at. Are the plugs at the meter end shrouded, so you can't accidentally touch bare metal? Are the handles on the probe flanged so that it is difficult to accidentally touch the bare metal? Both these are needed to help prevent you suffering a shock.

In summary, the most important criteria for choosing a meter are a clear, understandable scale, good basic ranges, and good test leads. Many other ranges and features are offered on multimeters; some are quite useful while others are simply a gimmick. The following facilities are not necessities but may occasionally be useful: battery test, transistor and diode test, and continuity buzzer.

CHAPTER 2

What is Electricity?

The atoms that make up the physical world are far too small to see, but scientists have built up a picture of what goes on inside them. Atoms consists of three sorts of particle (Fig. 2.1). *Neutrons* and *protons* are found at the centre of the atom, which is called the *nucleus*. Neutrons and protons are almost exactly the same weight and size, but protons have a positive electrical charge which makes them repel one another. Inside the nucleus, a very strong force – the nuclear force – overcomes the electrical repulsion and binds protons as well as neutrons together.

Forming a cloud around the nucleus are the negatively charged *electrons*, which are kept close to the nucleus by their attraction to the positively charged protons, but spread around in the cloud by their own mutual repulsion. Electrons are about one two-thousandth the weight of protons and neutrons. However, they have exactly the same size of electrical charge as protons, though negative. The simplest atom, that of hydrogen with just one proton in the nucleus and one electron, is electrically neutral.

The number of protons in an atom determines the kind of atom it is. An atom with eight protons is always oxygen. The number of neutrons is

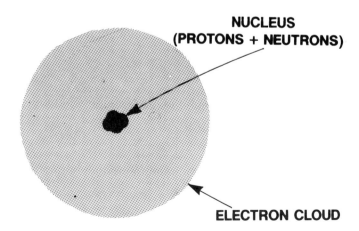

Fig. 2.1 A diagram of an atom.

usually around the same as the number of protons in lighter nuclei, more in heavier. Oxygen usually has eight neutrons, but a very few oxygen nuclei have nine or even ten. Uranium has 92 protons, and usually either 143 or 146 neutrons.

The nucleus attracts electrons to it to balance out its positive charge. These electrons form distinct *shells* within the cloud. The inner shells are bound closely to the nucleus but, in the outer shell, electrons can be shared or exchanged with other atoms, and by doing this the atoms bond together. This outermost shell is called the *valence shell*.

Elements are substances which are composed of atoms of just one type; *compounds* are substances which have more than one sort of atom in them. For example, water is composed of two atoms of hydrogen bonded to one atom of oxygen to form a molecule, which is defined as the smallest particle of a substance which keeps the properties of that substance.

Very few materials in common use are composed of just one element. One is copper wire, which is pure copper. But wood, for example, consists of a very large number of different molecules.

Conductors and current

In some substances, the bonding of atoms to one another leaves some electrons (usually the valence electrons) free to move around. Any flow of electrons is called a *current* and a substance which allows a current to flow easily is called a *conductor*. All metals are conductors, and the best conductor (the one in which electrons can move most easily) is gold. Gold is expensive, so we usually use copper which is nearly as good. When we refer to 'a conductor' meaning an object, we nearly always mean a length of copper wire.

There was a gap of thousands of years between the discovery of electricity and its scientific explanation, and in the meanwhile it had been decided to think of electricity flowing from positive to negative. This is known as the *conventional direction* of current flow, and we also refer to *conventional current*. However, because electrons are negative, they actually flow from negative to positive (Fig. 2.2). Fortunately, this rarely makes any difference.

The unit of electrical current is the *ampère*, usually abbreviated to *amp* or the symbol A. One amp of current is flowing when 6,241,807,000,000,000,000 electrons pass a point every second (about six million million million electrons). An amp is quite a large current in electronics, and we frequently use milliamps (mA) and microamps (μA or uA, where μ is the Greek letter mu). 1000 microamps equal one milliamp, and 1000 milliamps equal one amp, so there are a million microamps in an amp. Also used are nanoamps (nA) and picoamps (pA). One nanoamp is a thousandth of a microamp and one picoamp is a thousandth of a nanoamp.

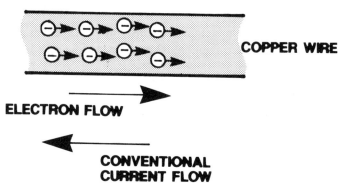

Fig. 2.2 Electrons flowing in a wire and the direction of conventional current flow.

The reducing prefixes milli, micro, nano and pico are used frequently, and so are the multipliers of kilo (k, x 1,000) and mega (M, x 1,000,000), so make yourself familiar with them. (A kiloamp is 1,000 amps and a mega-amp is 1,000,000 is a million amps – a very large current indeed, perhaps the size of current in a lightning strike.)

The symbol I is used to describe current when writing formulae; to distinguish different currents, we add numbers or letter subscripts, for example $I2$, I_B, and so on. A collection of electrons represents a certain amount of *negative electric charge*; alternatively, a collection of atoms which are short of electrons represents a *positive charge*. The unit of charge is the *coulomb*, which is the amount of charge which has passed along a wire if a current of one amp flows for one second. We could equate this to six million million million electrons or the same number of atoms with one electron missing.

Insulators and semiconductors

In some substances, the electrons are not free to move so current cannot flow. These substances are called *insulators*. Many substances when pure and dry are insulators, and plastics in particular are generally very good insulators.

In between insulators and conductors are *semiconductors*, which are substances that allow electrons to move round them but which themselves, when absolutely pure, have no free electrons. Very interesting and useful behaviour can be obtained by adding small quantities of other, precisely selected atoms. Semiconductors are described in Chapter 4.

EMF, PD and volts

For an electric current to flow, there must be a pushing force to make the

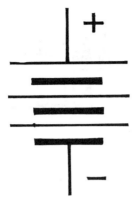

Fig. 2.3 The circuit symbol for a battery.

electrons move. This pushing force can come from a number of sources, most often either chemical (batteries) or from electromagnetism (electricity and magnetism together, as in generators, dynamos and alternators). For the purpose of this book, there is no need to go into how these work – you will find explanations in books on electricity.

The pushing force from a source is known as the electromotive force, or *EMF*. This term is used very strictly for the primary pushing force from a battery or generator (although we also talk of a 'back-EMF' from some magnetic components, where some but not all of the components of a generator are being used). Once we are away from these, we use another term. *Potential difference* or *PD*, is the correct term for the pushing force, but *voltage* is almost universally used instead because volts are the units which are used to measure both EMF and PD. Both PD and the unit volts have the symbol V, so you have to be careful to distinguish between them.

An ideal battery (or generator) would have the same pushing force whatever was connected to it, and the potential difference on the battery's terminals would always equal its EMF. However, drawing a heavy current from the battery causes the potential difference to drop and we speak of the *terminal potential difference* to distinguish this from the EMF. We think of there being an ideal battery inside which always has the same EMF, but which has an imperfect connection with its terminals.

Resistance

If we were to take a piece of carbon and apply a voltage across it, we would get a current flowing through the carbon. If we vary the voltage and measure both the voltage and the current, we will find that they go in step with each other. There is a constant factor linking them together called the *resistance*, which is a property of that particular piece of carbon.

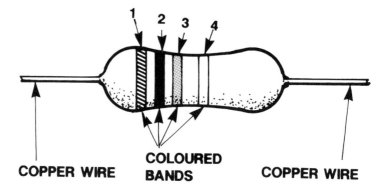

Fig. 2.4 A typical small resistor, showing the colour coded bands which are used to mark the value of the resistor.

Mathematically, the voltage divided by the current is always equal to the resistance. Using the symbol R for resistance, we can write three different version of the formulae for this, and use whichever version is most convenient:

$$R = \frac{V}{I}$$
$$V = R \times I$$
$$I = \frac{V}{R}$$

This rule is known as Ohm's Law, and it is the single most useful rule in electronics.

The unit for resistance is the ohm, symbol Ω (Greek omega). One is the resistance which allows a current of one amp to flow through it when there is a voltage of one amp across it. In electronics, resistances of kilohms (kΩ) and megohms (MΩ) are encountered. $1000\Omega = 1\text{k}\Omega$ and $1000\text{k}\Omega = 1\text{M}\Omega$, so that $1{,}000{,}000\Omega = 1\text{M}\Omega$. It is also quite common to leave out the symbol because many typesetters or typewriters cannot provide it. R is used for ohms, so 100R means 100Ω.

We can now be more precise over what constitutes a good or poor conductor – it is one with a very low resistance. All conductors have some resistance, and though it can be ignored most of the time, it should never be forgotten.

Resistors

These are one of the commonest electronic components, and they are devices which are made to have a specific, predetermined resistance, usually marked on the resistor's body using a colour code.

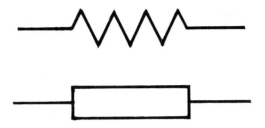

Fig. 2.5 The two circuit symbols in common use for the resistor; I prefer the top one.

There are series of resistance values that are commonly available, the most common being known as the E12 series: these are 10, 12, 15, 18, 22, 27, 33, 39, 42, 47, 56, 68 and 82 ohms, and values 10, 100, 1000 and 10,000 times these. These values may seem odd at first sight, but they are designed to have a ratio of close to 1.212 (the 12th root of 10) between adjacent numbers, i.e. 15:18 is about the same as 47:56, and so on. For very large and very small values, availability is usually limited to the E6 series, which has half as many values. Some suppliers may carry the E24 range, which has extra values in between the E12 values. Table 2.1 lists all the E6, E12 and E24 values. There are also E48 and E96 series, but these are very rarely used.

There are two ways of writing resistor values, either with a conventional decimal point or with the multiplier (R for ohms, k for kilohms or M for megohms). 4.7 ohms can be written as 4R7, 1.0 kilohms as 1k0 and 1.5 megohms as 1M5. In electronics, we usually use the multiplier, i.e. 4R7, 1k0 and 1M5.

Originally, resistors were made either of solid pieces of carbon or from lengths of special resistance wire wound on a former. Solid carbon resistors have all but disappeared, being replaced by ones made of films deposited onto usually ceramic formers or cores; the film itself can be carbon, metal or metal oxide. Wire wound resistors are still used where there is a high power dissipation – this is explained later on. The *tolerance* of a resistor is a measure of how accurately it has been made; so a 100 ohm resistor with a 5% tolerance can be any value between 95 and 105 ohms. The lower the tolerance of a resistor, the more expensive it is, but tolerances of 2% and even 1% are becoming increasingly usual as manufacturing methods improve. The value and tolerance of the resistor is normally marked on the body using a colour-code (most resistors are too small to have the value written on them). Table 2.2 shows the code used and, with practice, you will find it easier to remember this table.

Examples using the colour code
red-red-red-gold = 22 × 100 Ω (2k2) 5%
brown-black-blue = 10 × 1,000,000 Ω (10M) 20%
yellow-purple-orange-red = 47 × 1,000 Ω (47k) 2%

Table 2.1 E6, E12 and E24 values. Note that they are not evenly spaced; however, the ratio between successive values in each series is always approximately the same.

E6	E12	E24
10	10	10
		11
	12	12
		13
15	15	15
		16
	18	18
		20
22	22	22
		24
	27	27
		30
33	33	33
		36
	39	39
		43
47	47	47
		51
	56	56
		62
68	68	68
		75
	82	82
		91
100	100	100

Note that the values are not evenly spaced; however, the ratio between successive values in each series is always approximately the same.

Combining resistors

What difference does putting two resistors in series make? The same current has to flow through both of them (Fig. 2.6) as the electrons entering the junction from one resistor's lead must all go into the other resistor's lead; let's call the current I. The voltage across the two together must equal the voltage across one, V1, plus the voltage across the other, V2. Putting this into maths:

$$V = V1 + V2$$

Table 2.2 Resistance colour codes. No tolerance band = 20% (rarely found except on very old equipment).

Colour	Band 1 First figure	Band 2 Second figure	Band 3 Multiplier	Band 4 Tolerance
Black	–	0	1	–
Brown	1	1	10	1%
Red	2	2	100	2%
Orange	3	3	1,000	–
Yellow	4	4	10,000	–
Green	5	5	100,000	–
Blue	6	6	1,000,000	–
Purple	7	7	–	–
Grey	8	8	–	–
White	9	9	–	–
Gold	–	–	0.1	5%
Silver	–	–	0.01	10%

No tolerance band = 20% (rarely found except on very old equipment).

If we call the resistors' resistances R1 and R2, we can use Ohm's Law on the voltages and current:

$$V1 = I \times R1$$
$$V2 = I \times R2$$

Putting these into the sum for V:

$$V = I \times R1 + I \times R2$$

which can be rewritten as:

$$V = I \times (R1 + R2)$$

Comparing this to Ohm's Law, we see that the resistance of the two resistors is equal to the sum of their resistances.

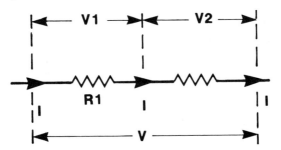

Fig. 2.6 Two resistors in series.

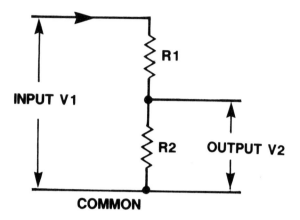

Fig. 2.7 Voltage divider circuit.

Using resistors to divide voltages

Fig. 2.7 is Fig. 2.6 redrawn with a common line added to make it a *voltage divider* (also called a *potential divider*). The circuit divides the input voltage by an amount set by the resistor values.

The dividing factor of the circuit can be found from Ohm's Law. The combined resistance of the two resistors is R1+R2, so the current flowing through both resistors is $V_{IN}/(R1+R2)$, where V_{IN} is the input voltage. The output voltage is R2 times the current flowing through it, which is $R2 \times V_{IN}/(R1+R2)$. The circuit has made the input voltage smaller by a factor of R2/(R1+R2). We can set any dividing factor we like by altering the relative values of R1 and R2, but we can never make the output larger than the input.

A controllable divider—the potentiometer

The voltage divider circuit is even more flexible when we can adjust it as needed. The *potentiometer* (or 'pot') does this. It is a single resistor (the track) with three connections, two at either end and one (the wiper) which can be varied in position along the track. The construction of a rotary potentiometer is shown in Fig. 2.8; another less common sort is a linear potentiometer, which has a straight track and a slider, but which otherwise works in the same way.

The resistor in a potentiometer can be made from any of a number of materials, the most common being carbon, special resistive ceramics and resistance wire. Standard potentiometers have an easily accessible shaft normally with a knob attached. Preset potentiometers are intended to be

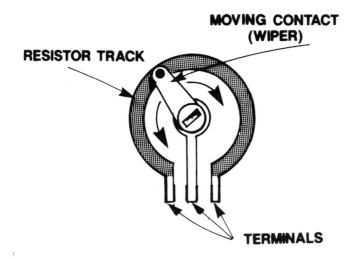

Fig. 2.8 Construction of potentiometers.

adjusted infrequently and a special adjusting tool (or small screwdriver) is needed.

One of the most common uses for potentiometers is for the volume control on radios, televisions (without remote control) and hi-fi amplifiers, where they adjust the size of a signal. For these jobs, we often use logarithmic potentiometers, where the resistance of the track is spread unevenly with a heavy concentration towards one end; this makes the sound level vary unevenly as the control is rotated, but this compensates for

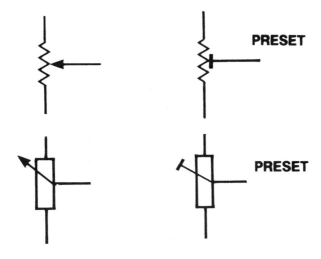

Fig. 2.9 The symbols for potentiometers; the ones to the right are used to indicate preset types.

the unevenness in human hearing. Our ears are proportionately less sensitive to loud sounds – a noise which sounds twice as loud as another is actually ten times or more larger, and we can barely distinguish the difference between sounds which are actually twice as strong. Most other applications use linear potentiometers, where the resistance is spread evenly along the track.

Resistors in parallel

The second way of combining resistors is in parallel (Fig. 2.10). The voltage V is the same across the two resistors, but the currents will be different. They can be determined by applying Ohm's Law to each resistor:

$$I1 = \frac{V}{R1} \qquad I2 = \frac{V}{R2}$$

The two currents add together to make the total current through the pair of resistors:

$$I = I1 + I2$$

If the pair of resistors has an effective resistance R, then using Ohm's Law again:

$$I = \frac{V}{R}$$

The previous equation can be written again but using voltages and resistances:

$$\frac{V}{R} = \frac{V}{R1} + \frac{V}{R2}$$

$$\frac{1}{R} = \frac{1}{R1} + \frac{1}{R2}$$

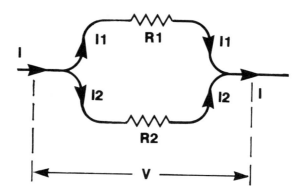

Fig. 2.10 Two resistors in parallel.

Since all the terms in this equation involve dividing into V, the second version has eliminated V. Another way of writing this is:

$$R = \frac{R1 \times R2}{R1 + R2}$$

I will leave it to algebraically-minded readers to show how this second version is arrived at from the first.

If two resistors of the same values are placed in parallel, the effective resistance is one half of their single values. If N resistors of exactly the same value (where N is any whole positive number) are placed in parallel, the effective resistance will be their individual value divided by N.

Power

When a resistor has a voltage across it, and so carries a current, heat is generated in it. The element of an electric heater is a resistor and, due to a combination of current flowing through it and voltage across it, the element gets very hot.

It is the combination of voltage and current which is important. The cable connecting the heater to the electricity supply does not get hot (or if it does, it needs replacing urgently) because individual conductors in the cable have negligible voltage across them. Between the conductors is the insulation, but this does not get hot despite the voltage across it – no current is flowing through it, it is all flowing in the conductors.

The power is the measure of the rate at which heat is being generated, and is measured in *watts* (symbol W): one joule of heat energy is generated in one second by one watt of power (from Physics, we know that 4.2 joules is the amount of energy needed to heat one gram of water by 1°C). There is a simple relation between power P, voltage V and current I for a resistor:

$$P = V \times I$$

The size of the volt was fixed to create this neat relationship, which applies to all electronics and electricity, and in alternating currents with certain reservations.

Two other ways of writing this formula can be used when we are dealing with just resistors (the main way can be used for anything):

$$\begin{aligned} P &= V \times I \\ &= I^2 \times R \\ &= V^2/R \end{aligned}$$

If we are supplying V volts and I amps to some electrical or electronic equipment, P watts of power (given by the first version of the formula above) must be coming out somewhere. The most usual way is in heat, but other forms of energy – sound, mechanical power (a motor moving something), storage in a battery or capacitor, radio waves–might be involved. The total is always the same–power in equals power out, provided any storage is taken into account.

Returning to resistors, practical resistors all have a power rating in watts which is the power dissipation that resistor can stand. If this power rating is exceeded for more than a short time, the heat generated will cause the resistor to burn up. The same limitation applies to nearly all the other components we will meet later.

Switches

Switches connect or disconnect the flow of electric current. The number of lines the switches can connect or interrupt simultaneously is the number of *poles* of the switch, so a three-pole switch is one which can connect or disconnect three sets of lines simultaneously.

In electronics, many switches can connect current from a single wiper line to any one of several *ways*. A single way switch can just connect or disconnect a single line; a four-way switch can connect the wiper to any one of four other lines. The number of poles and the number of ways need to be specified. If you need to switch the mains electricity supply, for example to make an on/off switch for a project, you must always use a specially designed switch. It is very dangerous to use a switch which is not specially designed.

Types of switch

Toggle switches have an operating lever and a spring mechanism which moves the contacts very quickly from one position to another, giving a

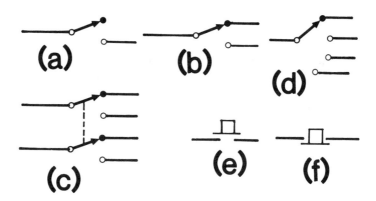

Fig. 2.11 Symbols for various types of switch: (a) single pole, single way; (b) single pole, two way; (c) two pole, two way or 2p 2w; (d) single pole four way; (e) a push-to-make switch; and (f) a push-to-break switch.

good switching action. Toggle switches can switch fairly high currents, and are often used for switching the mains supply to equipment.

Rocker switches are operated from a large rocker on top, and they too are often used for switching the mains.

Rotary switches are controlled by turning a knob, and are often used for wave-band switches on older radios. Rotary switches are very adaptable. Typically, you can get standard switches of 1 pole 12 way, 2p 6w, 3p 4w, and 4p 3w, as well as kits to make many more combinations. Although most rotary switches cannot switch the mains, special mains types are available.

Push buttons are everywhere. Some are single pole which make their contact only while they are held in. With others, pushing makes the contact and pushing again breaks it. Yet others are mounted in rows using a special bracket: when one is pushed in, the others pop out. Again, special mains switches are available.

Slide switches consist of a sliding plastic knob attached to contacts, which are usually one or two pole, two way; they do not have a good switching action, but are very cheap. They are not suitable for the mains.

Micro-switches are misnamed by today's standards, as many of them are fairly large. They are designed to be position sensors, operated by a lever.

DIL switches are miniature toggle or slider switches mounted in rows, in a single body; these are used often for computers and peripherals (DIL stands for dual in line, which refers to the connector pins of the switches).

Membrane switches are simplified push switches, covered by a continuous membrane which seals them from liquids; these are used on equipment such as drinks vending machines and photocopiers.

All switches have contact ratings which give the maximum voltage and the maximum current they can switch without damage. Often there will be two sets of ratings, for resistive and inductive loads. A resistive load is something simple like a light bulb or an electric fire; inductive loads are described later, but in this case it means virtually anything else. Include anything you are not sure about in this category.

The contacts of switches with more than one throw can either be make-before-break, or break-before-make. In a make-before-break, the contact that is being disconnected from the wiper is momentarily connected via the wiper itself to the contact that is being made. Sometimes this is useful, at other times it can cause disaster. Unless specified otherwise, always use break-before-make.

Fuses

The job of a fuse is to 'blow', that is permanently break the contact between the two ends, when the current through it exceeds the value marked on the fuse's case. Electronic fuses are either 20mm or $1\frac{1}{4}$ inches in length, and they are available in a much wider range of currents than normal household fuses. There are three different types.

Delayed fuses (also called 'slow-blow', T-type, anti-surge and semi-delay) can withstand for a short time a surge current which is several times the normal fuse rating. If the current does not fall to below the fuse's rated current in a tenth of a second or so, they will blow. These fuses are used in things like power supplies which normally draw a high current for a fraction of a second when first switched on.

Fast fuses (also called quick-acting, 'quick-blow' and F type fuses) are the type usually used in electronic equipment away from power supplies. They are made to blow quickly if the current goes above their maximum. There are also *super-fast fuses* (or super-quick acting or FF type fuses), which are even faster and are used to protect vulnerable equipment.

On any sort of equipment, if a fuse keeps blowing, never replace it with a higher value fuse (unless the original fuse was of the wrong rating). If a correct fuse blows more than twice in quick succession, look for the reason before replacing it again.

Getting projects built

There are several methods for assembling electronic projects. *Birds-nesting* aptly describes what happens when you just solder components directly together using only their leads. It is fine for trying out small circuits, but it is very vulnerable to leads shorting together and unsuitable for permanent items.

Breadboards have rows of sockets connected together which component leads are pushed into. You can try out complicated circuits on a breadboard, but it is not suitable for permanent items.

Strip-board consists of a board with a matrix of holes through it, with strips of copper on one side joining up the holes in each row. It is like a permanent version of a breadboard. Components are placed with their leads through the holes and soldered to the copper strips. Strip-board is suitable for permanent items, provided care is taken and both the current and voltage on the board is quite low; it is not recommended for circuits which use the mains. Quite sophisticated versions of strip-boards are available.

Printed circuit boards (PCBs) are custom versions of strip-board. The components are soldered into the holes in the board and specially designed copper tracks make the circuit connections. All the PCBs used in this book are single-sided, with the copper tracks on the reverse side. Double-sided boards are common, and multi-layer boards with several layers of tracks, separated by insulating layers, are used in industry.

Other methods also exist. In the Speaker Splitter project, connections are

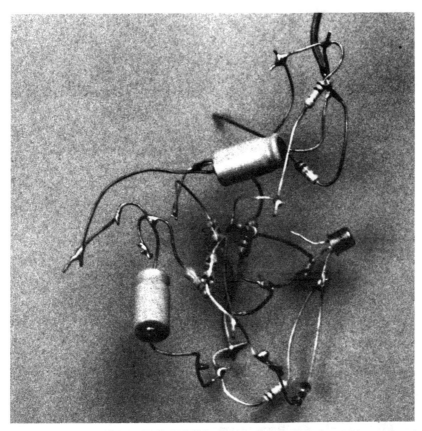

Fig. 2.12 Bird's nesting (also called rat's nesting, for similar reasons) is rough and ready but liable to give unwanted connections.

Fig. 2.13 Breadboards are great for try-outs, and are designed to be re-used again and again, so they're not suitable for permanent circuits.

Fig. 2.14 Strip-board (often called Veroboard after the leading brand) is great for one-offs and experiments.

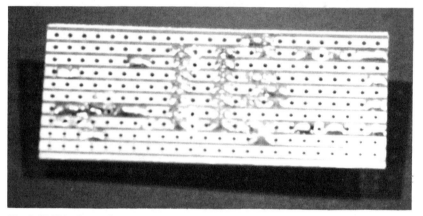

Fig. 2.15 This shows the underside of strip-board, where you cut tracks with a drill for special tool to stop connections you don't want.

made using connector blocks. In older TVs and radios, tag-strips are used to make a permanent version of birds-nesting.

Soldering

This is one of the basic skills in electronics, and it is worth practising before trying to solder a 'real' circuit.

Solder is a mixture of lead and tin, in electronics the mix is 60% tin and 40% lead which makes it go straight from liquid to solid as it cools. It is sold in reels of 'wire' with cores of flux running through the middle. Flux is a chemical cleaner which scours the surfaces to be joined, making it easier to get a good joint; it only works when heated up by the soldering iron. The melting point of solder is 188°C, which is high enough to damage many electronic components if their leads are kept at this temperature for too long. It is well worth practising soldering, so buy a 'bargain pack' of mixed components and some 'offcuts' of strip-board.

Place a component so that its body is on the top of the board and its leads poke through to the copper strip side. Bending the leads slightly will keep the component in place. Allow the soldering iron to heat up fully, then tin the bit by applying a little solder to it – there will be a slight hiss and some fumes given off as the flux in the solder scours the surface of the bit. Solder the component leads to the copper strips by applying the hot bit and the solder simultaneously at the point where the joint should be. Feed in enough solder to make the joint look neat, then firstly remove the solder and secondly the bit.

Of course, it won't be easy at first, but with repetition you will find it gets much easier to make neat joints. A typical joint should take a few seconds: any faster and you have probably got a dry joint (one where the flux has not had time to do its job); much longer and you risk damaging the components. Try soldering leads in adjacent holes that are not joined by the same copper track. If you get a bridge of solder between the two tracks – and you will every now and then – use a solder sucker de-soldering tool, de-soldering braid or some bare stranded wire to mop up the excess and try again.

Some books advise using crocodile clips or special heat sinks to prevent damage to components but I don't know of anyone who uses them. Electronic components are designed to withstand soldering if it is done efficiently.

CHAPTER 3

When the Voltage Keeps Changing

So far we have been dealing with direct voltages and currents which keep the same direction. Alternating voltages and currents are different, as they constantly change direction. Because they can drive transformers (devices we will meet later in this chapter), alternating voltages are used for the mains electricity nearly everywhere in the world.

The widely used abbreviation 'AC' means alternating current; 'DC' means direct current. The terms 'AC voltage' or 'DC current' are slightly nonsensical, but AC and DC are used almost universally as shorthand for 'alternating' or 'direct'. The most common alternating voltages and current do not change direction at random, but regularly and smoothly. Fig. 3.1 shows such an alternating voltage over a short period of time. This shape, known as a *sine wave*, is also one that alternating current normally has. Starting from zero, the voltage increases quickly, but the rate of increase gradually slows until a peak is reached. After this, the voltage decreases slowly at first then increasingly quickly until it again reaches zero. It then carries on getting more negative, but at decreasing speed until it slows to a stop at the negative peak – here it turns back again and gets less negative increasingly quickly until it again reaches the zero line; after this the whole waveform repeats.

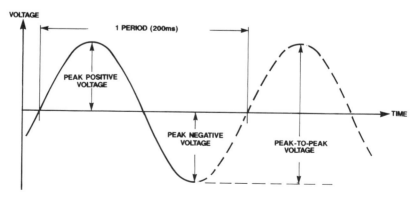

Fig. 3.1 The shape of a sine wave.

For any sine wave, the positive peak voltage and the negative peak voltage are equal in size though opposite in direction; the *peak voltage* or *amplitude* (symbol V_p) means the size of either. The *peak-to-peak* voltage is the difference between the two peaks, twice the peak voltage.

A complete wave is known as a *cycle*. The cycle can be measured from any two corresponding points on the wave, but it is convenient to measure between two zero-crossings in the same direction, as Fig. 3.1 does. The period of this sine wave is a fifth of a second, 200 milliseconds. Five cycles of this sine wave would occupy a second exactly: its frequency – the rate at which it repeats – is 5 cycles per second, or 5 hertz (pronounced, 'hurts', symbol Hz).

RMS Voltage

We need to know what effect an alternating supply will have, for example how much heat it will generate in an electric heater; we need to work out the average power – what V and what I should be used in the power formula? The peak values are appropriate only at the peaks themselves.

As V and I change together, it will be easier to think in terms of just one or the other; let us use $P = V^2/R$. Since R can be found by measuring the resistance of the element, all we need is the average (or mean, to be more precise) of V^2. A characteristic of sine waves is that whatever their size, the mean of V^2 is always equal to half the square of the peak voltage, $V_p{}^2$. This is not true for other possible wave shapes.

If we take square roots, V becomes the root-mean-square voltage, V_{RMS}, which is equal to V_P divided by the square root of 2, 1.414. V_{RMS} is not the same as the average of the sine wave's voltage, which would be zero, nor is it the average of the positive section of the sine wave. V_{RMS} can be used in all the same formulae as DC voltages, including the power formula $P = V \times I$, provided that the RMS current also is used (it is the peak current divided by square root of 2). So useful is the RMS voltage that alternating supplies are always specified in volts RMS. In the UK, the mains voltage is 240V RMS and the peak voltage is $240 \times 1.414 = 339V$. In the USA, the mains is 110V RMS so the peak is 156V. If a 100Ω resistor is connected across the mains supply in the UK, the power dissipated will be $240^2/100 = 574W$, averaged over the whole cycle of the sine wave.

In electronics we encounter many other types of wave – for example, the signal in a hi-fi – and waves may or may not repeat regularly. However, most can be characterised by their RMS voltage, which depends on the wave shape and the peak value. Knowing the RMS value allows us to calculate how much power these waves will dissipate in loads, for instance, in a loudspeaker.

Frequency

The frequency of sounds that the human ear can hear extends from 20Hz to 20,000 Hz (20 kilohertz, kHz). Alternating signals in this range are often called audio frequency or AF signals. Radio frequency (RF) signals usually have frequencies between 100kHz and 1GHz (1GHz = 1,000,000,000Hz). RF signals and radio waves are different. RF signals move along a conductor whereas, radio waves are akin to light waves and need no conductor. RF signals can be converted into radio waves (or vice-versa) by an antenna (or aerial) which is a transducer – it turns one form of energy into another.

Real signals

Most signals are a mass of jumbled, individually repeating and non-repeating waves. Consider the output from an electric guitar pick-up when a single string is struck; the signal will be a steadily repeating signal, which dies away gradually, plus some non-repeating parts when the string is first hit with the plectrum. Looking just at the repeating parts, these can be broken down into a number of sine waves. The lowest in frequency is called the fundamental (this is the pitch of the note). All the other sine waves have frequencies which are multiples of the fundamental; they are called harmonics or overtones.

Any signal which repeats regularly can be broken down this way, or built up from sine waves of the right frequency and size. Many electronic circuits react to complex waveforms as a collection of sine waves.

Charge and capacitors

Capacitors store electric charge, and capacitance is what we call the property that allows them to do this. A very basic capacitor (Fig. 3.2) consists of two metal plates separated by a layer of perfect insulator called the dielectric; leads are attached to the plates for external connections. If this capacitor is connected across a battery, an excess of electrons (negative charge) builds up on the negative plate while a deficit of electrons (positive charge) builds up on the positive plate. The higher the voltage of the battery, the larger the charges on the plates. If the connections to the battery are removed, the charges on the plates and the voltage between them remain the same (the capacitor is charged). In practice, all dielectrics conduct slightly and the charge leaks away slowly.

Take a 100 microfarad capacitor (with a voltage rating of at least 10 volts), a multimeter and 9V battery (Fig. 3.3). With the meter on a 10V (or higher) DC volts range, attach the leads across the capacitor as shown using clips. Then touch the positive lead of the capacitor to the positive

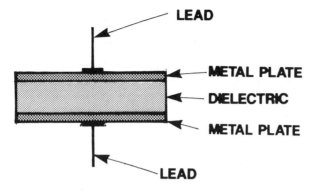

Fig. 3.2 A very simple capacitor.

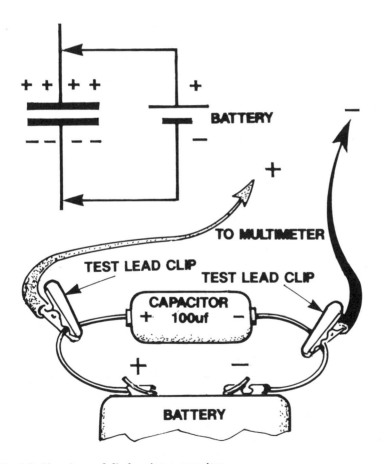

Fig. 3.3 Charging and discharging a capacitor.

battery terminal and the negative lead to the negative terminal (be careful to get these the right way round), then remove the battery. Before reading on, try to explain the meter reading you see.

The multimeter needle will go to around 9V (depending on the battery's health), then it will fall back slowly towards zero after the battery is disconnected (there may have been a small spark when the capacitor was first connected to the battery – this will be explained shortly). The battery charges the capacitor up to its own voltage, but after it is disconnected the capacitor's charge leaks slowly away through the multimeter.

Charge is measured in units of coulombs. One coulomb is the amount of charge in one amp flowing for one second. It can be negative (electrons) or positive (lack of electrons). Capacitors always store equal amounts of positive and negative charge. We measure capacitance in units of Farads (symbol F). A capacitor of 1 farad can store 1 coulomb of charge with a voltage of one volt across it, 2 coulombs with 2 volts, and so on. One farad is a very large capacitance; capacitors of this size have only recently become widely available. Microfarads (μF) and picofarads are much more commonly used. When the 100μF capacitor was connected across the 9V battery, the charge on it was $9 \times 100/1,000,000$ coulombs or 900μC (microcoulombs).

In summary, when capacitors are connected across a steady voltage, they store charge on their plates; how much charge depends on the sizes of the voltage and the capacitor. When the voltage is steady, the charge is steady and no current flows in or out of the capacitor. If the voltage is changed suddenly, a large current can flow, hence the slight spark when the battery is first connected.

Capacitors, alternating voltages and reactance

When a capacitor is connected across an alternating voltage, it will have charges constantly building up or decreasing on its plates depending on whether the voltage is increasing or decreasing. This means current has to flow into or out of the capacitor; the size of the current depends on how quickly the charge has to change, which in turn depends on the size of the alternating voltage, its frequency and the size of the capacitor (the higher any of these are, the larger the current).

The only time when current is not flowing is at the peak voltages, when the voltage momentarily turns from changing in one direction to changing in the other. Current flow will be highest when the voltage is crossing zero, because the rate of change of voltage is highest. The current and voltage will be out of step, as shown in Fig. 3.4. The current is a sine wave like the voltage, but it leads the voltage by a quarter of a cycle.

Reactance is the capacitor's equivalent of resistance, but it differs in two important ways: firstly it depends on the frequency of the applied voltage,

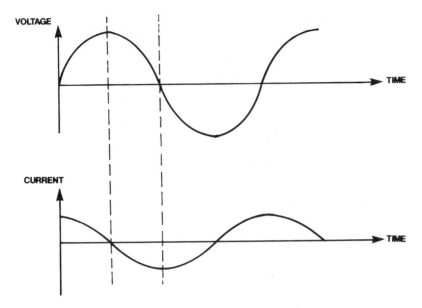

Fig. 3.4 Voltage and current in a capacitor across an alternating voltage.

and secondly no power dissipation occurs as a result of reactance alone because the current and voltage are out of step; the formula for power dissipated, $P = V \times I$, cannot be used. Besides capacitive reactance, there is inductive reactance which we will meet shortly. A formula can be used to calculate the size of a capacitor's reactance, Z_C:

$$Z_C = \frac{1}{2 \times \pi \times f \times C}$$

f is the frequency, C is the capacitance. The reactance of a $1\mu F$ capacitor at 50Hz is 3200Ω.

Real capacitors

Small value capacitors are made from thin foil plates in parallel, separated by thin layers of plastic, ceramic or other insulating material, as shown in Fig. 3.5. The more plates, the larger their area and the thinner the insulator the larger the capacitance. Often the foil plates and insulating films are rolled up to save space. Capacitors made this way have values of a few picofarads (pF) up to a few microfarads (μF).

To get higher values of capacitance, electrolysis is used to deposit a thin film of oxide onto a roll of foil immersed in a bath of chemicals (Fig. 3.6). The foil is one plate, and the chemical bath is the other 'plate'. The oxide film is very thin– perhaps only a few hundred molecules thick. Electrolytic

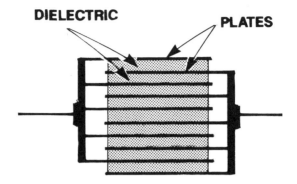

DIELECTRIC   **PLATES**

Fig. 3.5 A real multi-layer capacitor.

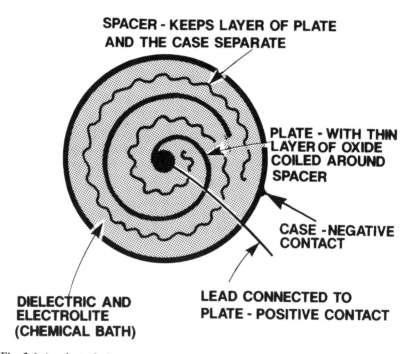

**SPACER - KEEPS LAYER OF PLATE
AND THE CASE SEPARATE**

**PLATE - WITH THIN
LAYER OF OXIDE
COILED AROUND
SPACER**

**CASE - NEGATIVE
CONTACT**

**DIELECTRIC AND
ELECTROLITE
(CHEMICAL BATH)**

**LEAD CONNECTED TO
PLATE - POSITIVE CONTACT**

Fig. 3.6 An electrolytic capacitor.

capacitors ('electrolytics') are polarised – they can withstand a voltage applied in one direction only. Applying a voltage in the other direction (or applying too high a one in the normal direction) breaks down the oxide film, destroying the capacitor. The capacitance of electrolytics varies according to the voltage applied (higher voltages make the oxide layer grow thicker) and they have relatively high leakage currents (they gradually discharge). Low leakage types exist, notably ones using tantalum in their chemical make-up ('tants') but these are likely to explode if connected the wrong way round or if they have too high a voltage applied. Special bipolar (also called non-polar) electrolytics can take voltages in either direction, but they are more expensive.

Capacitor values are marked either with numbers or with the resistor colour code (but with picofarads as the basic unit). The numbers can be either simply the value written normally, or a three number code where the first two numbers are as normal but the third is the power of the multiplier. For example, 334 means $33 \times 10^4\text{pF} = 330\text{nF}$. The maximum voltage that the capacitor can stand should also be marked, but often this is left off the smaller capacitors. Other capacitor characteristics can be important, and these are specified on manufacturers' data sheets. These include the inductance (see later) and the power factor. Ideally both these should be as low as possible, because both represent non-ideal characteristics. The power factor indicates the amount of power dissipated when a capacitor is connected across an alternating voltage (ideally, it would be zero).

Combining capacitors

What happens to the overall capacitance of capacitors when they are connected together? Because reactance relates the voltage and current, we can use the formulae for combining resistances with the reactances. With two or more capacitors C1, C2, C3 etc., in series, the overall reactance is given by:

$$X_T = X_{C1} + X_{C2} + X_{C3} + \ldots$$

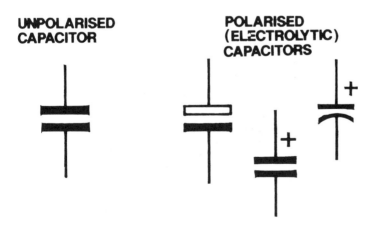

UNPOLARISED CAPACITOR

POLARISED (ELECTROLYTIC) CAPACITORS

Fig. 3.7 Capacitor symbols.

Using the formula for the reactance of capacitors and then eliminating common factors, this becomes:

$$\frac{1}{C} = \frac{1}{C1} + \frac{1}{C2} + \frac{1}{C3} + \ldots$$

For capacitors in parallel, the formula is:

$$\frac{1}{X_C} = \frac{1}{X_{C1}} + \frac{1}{X_{C2}} + \frac{1}{X_{C3}} + \ldots$$

But when the formula for reactance is used this simplifies:

$$C = C1 + C2 + C3 + \ldots$$

In this last result, we can think of capacitors' plates as all being connected. It is therefore quite natural that the overall capacitance should be equal to the sum of individual capacitances.

Inductors

Inductors are made by coiling wire around a plastic former or a core of magnetic material (Fig. 3.8). The number of times the wire is coiled around is known as the number of *turns*, and the wire itself is called a *winding*. When a steady current flows through the coil, it builds up a magnetic field around it; the more turns and the more magnetic the core, the stronger the magnetic field for a given current. Because an inductor is basically a length of wire, when a steady current flows there is no voltage across it (this assumes the resistance of the wire is too low to matter). If the current

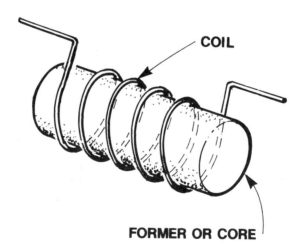

Fig. 3.8 A basic inductor.

changes, however, there will be an EMF induced across the inductor because the magnetic field is changing. The induced EMF opposes the change in current, and to get the current to change a voltage must be applied to overcome the induced EMF. The higher the rate of change of current, the higher the voltage induced, and the higher the applied voltage needed.

An inductor has an inductance of one *henry* (symbol H) if one volt is induced across it while the current is changed at a rate of one ampère per second. Although inductors of one henry or more have been available for some time, sizes of millihenries (mH) and microhenries (μH) are now more common. Inductors are also called coils and chokes (because they limit the speed at which current can change and so 'choke off' higher frequency signals).

Inductors and AC

When an alternating voltage is applied to an inductor, current flows but it lags the applied voltage by a quarter cycle (Fig. 3.9). The inductor has inductive reactance i.e. the voltage leads the current, unlike capacitive reactance where the current leads the voltage. As with capacitors, the reactance gives the ratio between current flowing and voltage applied,

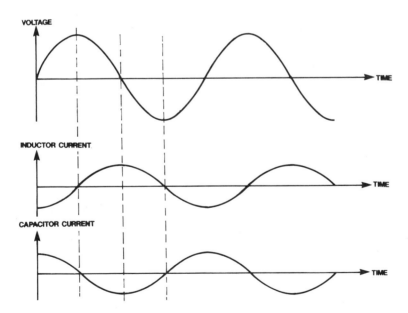

Fig. 3.9 The relationship between voltage and current for an inductor, and compared with a capacitor.

$V = X_L I$, where X_L is the reactance. The reactance depends on frequency and can be found from:

$$X_L = 2\pi fL$$

where f is the frequency in hertz and L is the inductance in henries; this gives reactance in ohms.

Like capacitors, ideally there is no power dissipation when an alternating voltage is applied, however, in practice, there is always some due to resistance in the windings, and this is particularly noticeable at high currents and low frequencies.

Combining inductors

Two or more inductors in series simply have the inductance of all the inductors added together, plus all their individual resistances added. There is no straightforward way to calculate the result of placing inductors in parallel because their resistances make the calculation very complex.

Combining resistors, capacitors and inductors

We cannot use any of the earlier formulae when combining different types of components. There are two ways of working out how resistors, capacitors and inductors will behave when combined. The first is to use phasor diagrams, and we will use this method on some of the simplest combinations; the second is to use complex numbers, but this is beyond our scope.

Impedance is the generalised term for resistance, capacitive reactance and inductive reactance; Z is its symbol. The impedance of a component (or group of components) governs the relationship between the voltage applied and the current passing through it. Besides the size of current flowing, it tells us whether it leads or lags the voltage and by how much.

Resistor and capacitor in series

If we apply an alternating voltage to a resistor and capacitor in series, what current will flow? First we draw an arrowed line labelled I to represent the current because although we do not know its size, it must be the same for both components (Fig. 3.10a). We draw another arrowed line labelled VR in parallel and next to the current line to represent the voltage across the resistor, which is in step with the current (Fig. 3.10b). From the top of the resistor's voltage line, we draw a line labelled VC at right angles to represent the voltage across the capacitor (Fig. 3.10c). The right angle signifies that the capacitor's voltage is a quarter-cycle behind the current. We draw a final line from the start of the V_R line to the end of the V_C line (Fig. 3.10d).

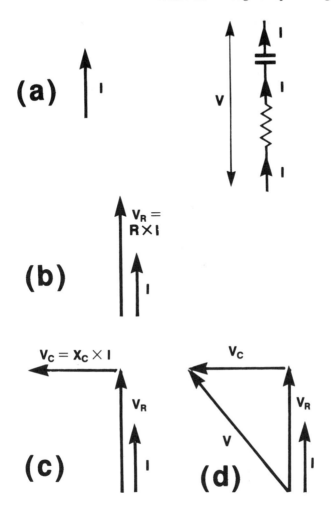

Fig. 3.10 Phasor diagrams for a resistor and capacitor in series.

Its length represents the total voltage and its angle to the current line represents how much the current leads the voltage.

V_C, V_R and V form a right-angled triangle, so we can use Pythagoras' theorem to find the size of V. (We also know that $V_R = R \times I$ and $V_C = X_C \times I = 2\pi fC \times I$.):

$$V = \sqrt{V_C^2 + V_I^2}$$
$$= \sqrt{(R \times I)^2 + (X_C \times I)^2}$$

Dividing throughout by I, we can find the size of the impedance:

$$\frac{V}{I} = Z = \sqrt{R^2 + X_C^2}$$

With the size of Z, we can find I by dividing V by Z.

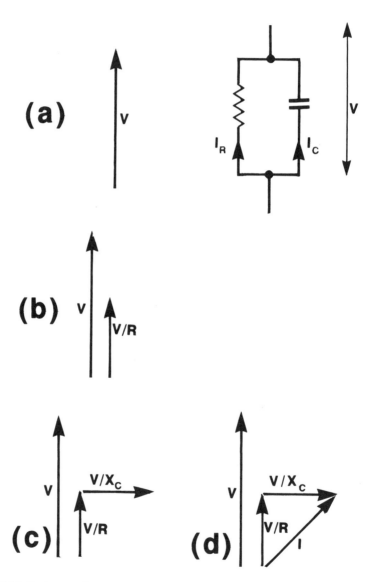

Fig. 3.11 Resistor and capacitor in parallel.

Resistor and capacitor in parallel

The same voltage is now across both components, so we can draw this first in Fig. 3.11a, then we can add a line for the current I_R in R (Fig. 3.11b), then a line at right angles for I_C through C (Fig. 3.11c). Completing the triangle gives the total current (Fig. 3.11d). Using Pythagoras' theorem on the currents:

$$I = \sqrt{I_R^2 + I_C^2}$$
$$= \sqrt{\left(\frac{V}{R}\right)^2 + \left(\frac{V}{X_C}\right)^2}$$

We can then convert the subject to Z, the impedance:

$$\frac{V}{I} = Z = \frac{X_C R}{\sqrt{X_C^2 + R^2}}$$

Resonance

When an inductor and a capacitor are in parallel and an alternating voltage is applied, their currents flow at exactly opposite times – the capacitor's current leads the voltage by a quarter-cycle while the inductor lags by a quarter. This is represented on the phasor diagram Fig. 3.12 by arrows for the current pointing in opposite directions. The capacitor's current goes up as frequency increases, while the inductor's goes down; at a particular frequency called the resonant frequency the two are the same and cancel each other out. Current flows around the circuit but none is drawn from outside; the circuit has infinite impedance.

With a capacitor and inductor in series, the current must be common so it is the voltages which now tend to cancel each other out (Fig. 3.13). At the resonant frequency, the two voltages cancel each other out and the circuit has zero impedance. In practice, there will be some resistance in the inductor, so the impedance will not quite be zero, but close. In both cases, resonance occurs when the reactance of the capacitor is the same as the reactance of the inductor. We can find the frequency from equating the two formulae. The result is:

$$f = \frac{1}{2\pi\sqrt{LC}}$$

Resonant circuits are often used in radio electronics to separate out wanted frequencies from unwanted ones. They are becoming more important in other areas too.

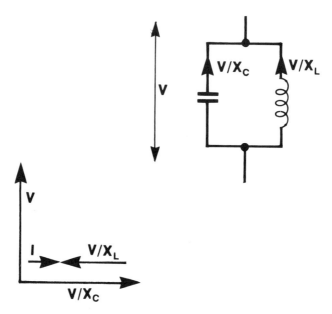

Fig. 3.12 Capacitor and inductor in parallel.

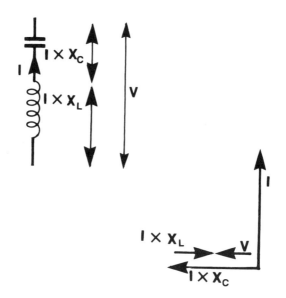

Fig. 3.13 Capacitor and inductor in series.

Setting limits

Fortunately it is rarely necessary to do calculations like the aforementioned. The main sort of circuit on which it can be necessary – filters – have already been analysed in great detail, and the results widely published.

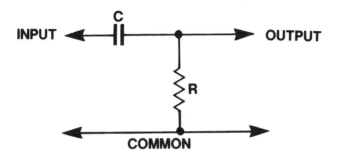

Fig. 3.14 Use of a coupling capacitor.

Often we have to decide whether a particular component is too large or too small, Fig. 3.14 demonstrates this point. C is a coupling capacitor and its job is to pass on an AC signal while blocking a DC voltage. The circuit's action is similar to a potential divider: for DC the capacitor's reactance is infinite, so no DC gets through. For all the AC signals of interest, the capacitor's reactance must be much smaller than the resistance, so no signal is lost. We have to find the minimum size for the capacitor to make this the case. Let us first find what value of capacitance makes the resistance and reactance equal. Equating the two:

$$R = X_C = \frac{1}{2\pi fC}$$

$$C = \frac{1}{2\pi fR}$$

If R is 1kΩ and the minimum frequency f is 20Hz – common values in audio signals – C must be at least 8μF. However, we will still lose a portion of the signal at 20Hz (although not quite half – see Fig. 3.10). To ensure only a small part of the signal is lost, let us make the capacitor much larger – 100μF is a common value we could use.

Transformers

Transformers are inductors with two or more windings (Fig. 3.15). Applying an alternating voltage to one winding, the *primary*, causes new voltages to be induced across other windings, called *secondaries*. These

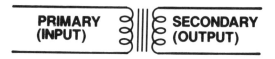

Fig. 3.15 Circuit symbol of a transformer.

secondary voltages are similar in origin to the EMF induced in a simple inductor. The size of the secondary voltage is the size of the alternating voltage applied to the primary, multiplied by the number of turns on the secondary divided by the number of turns in the primary. The secondary voltage can be larger or smaller than the primary voltage. For example, if a transformer has 1000 turns on the primary and 2000 on the secondary, the secondary voltage will be 2000 ÷ 1000 times or twice the primary voltage.

The two main uses for transformers are for converting the voltage of the mains (both in the supply companies' systems and in the home for domestic equipment), and in radio circuits.

Mains transformers

Mains transformers now usually supply output voltages much lower than the mains voltage (although in old equipment using valves the transformer may supply a higher voltage). They also isolate the rest of the equipment from the mains. They are specially designed to operate at the mains voltage and frequency, and it is dangerous to connect any other sort of transformer across the mains.

Ideally, there is no power loss in the transformer, so that the power in (volts times current can be used here) is equal to the power out; in practice there is usually some loss of power. Two types of mains transformer are in common use — laminated and toroidal. Laminated transformers are cheaper but tend to have higher losses than toroidal types. Fig. 3.16 shows the two types.

Each secondary has a fixed overall output voltage (since the primary voltage is fixed), but connections partway along called *taps* make it possible to use just part of the secondary. For example, one supplier lists a transformer with two secondaries as 0-10-12-15-17V, so any one of these voltages could be used on either secondary up to the full 17V. Another has its secondary described as 12-0-12V; this is a 24V secondary with the tap exactly halfway along (a *centre tap*); it can be used to supply either 24V or two linked 12V voltages. Each secondary also has a maximum current it can supply. Taps can also be used on primaries to allow the same transformer to operate from different input voltages (e.g. 110V in the USA, 220V on mainland Europe, and 240V in the UK).

Transformer secondaries can be connected together in series to add their

Fig. 3.16 Laminated and toroidal transformers.

voltages (but if connected the wrong way around their voltages will subtract). Two identical secondaries can be connected in parallel to double the available current.

PROJECT: Speaker splitter

Normally, if too many loudspeakers are connected to an amplifier, it will be overloaded because the speakers need more current than it can supply. The amplifier can be damaged even at low volume levels. With the speaker splitter, you can run three pairs of speakers from one stereo amplifier, but easily switch the amplifier to just the main speakers when required.

The splitter uses resistors in series with the individual speakers to increase the overall impedance of the loudspeakers and make it equal to the ideal impedance that the amplifier best drives. Fig. 3.17 shows the circuit of the splitter. The disadvantage of this circuit is that quite a lot of power is lost in the resistors, but if the loudspeakers are used for low-level background music, this will not be a problem.

We can think of a loudspeaker's impedance as 8Ω resistive for our purposes, although this is a very large approximation. Looking at R1 and loudspeaker LS1 on the circuit, these are in series and so their impedances add together; the effective impedance is $R + 8\Omega$, where R1, R2 and R3 all have the same value R. There are three groups of resistors and loudspeakers in parallel; the effective resistance is $(R + 8)/3\Omega$. Making R equal to 16Ω would make the overall resistance 8 ohms. Actually, 16Ω is not a widely available value, but 15Ω is and this will do. Switch SW1 allows all the power from the amplifier to be diverted to just the main speakers.

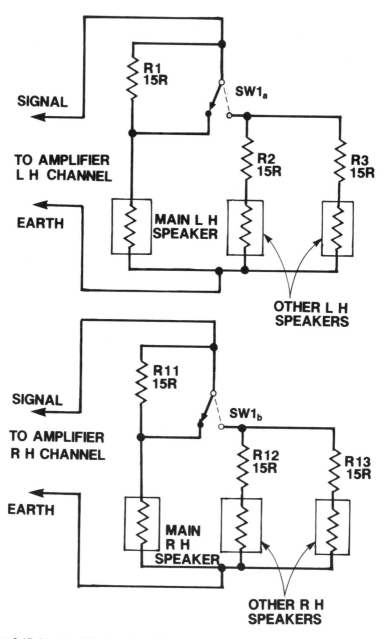

Fig. 3.17 Circuit of the Speaker Splitter.

Note that on the circuit diagram, all the components for the right-hand stereo channel have the same numbering as the left-hand side but with 10 added to their number (except for SW1b, which is part of the same switch as SW1a).

Putting it together

Planning is vital to success in any project, so make sure you understand how it will all fit together before you begin. The parts for this project are listed in the parts list, although the wire needed to connect the extra loudspeakers and the speakers themselves are not included. You will need a selection of small tools, including a soldering iron and an electric drill. Use the drill to make a hole for the switch in the front of the box; the front panel can be detached on some boxes which makes drilling much easier. A small hole may be needed for the locating lug which some switches have to stop them accidentally twisting round. Also drill 8 holes in the rear of the box for the loudspeakers' leads.

Solder three 100mm lengths of wire with different colours to the solder tags on each side of the switch (use the same colours for corresponding tags on the two sides). First strip both ends of the wires and tin one end of each with some solder (hold the wires in a vice or some pliers with an elastic band round the handles so your hands are free for the solder and iron). Then hold switch SW1 gently in the vice or pliers, and tin its tags. Finally attach the wires to the switch tags using the soldering iron. No more solder should be

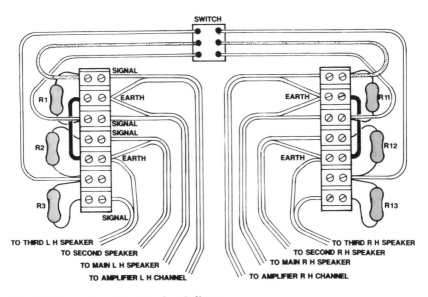

Fig. 3.18 Assembly of the Speaker Splitter.

Fig. 3.19 The inside of the Speaker Splitter is pretty crowded!

Fig. 3.20 The Speaker Splitter in use with the loudspeakers built in Chapter 7.

needed, but watch out for the wires getting hot. Make sure you don't leave any solder bridges between the tags.

Put 20mm of sleeving (the covering from stranded wires is a good substitute) on each of the resistor's leads, pushing it right up to their bodies. Cut the ends of the leads so that about 5mm remains exposed. Circuit assembly is as shown in Fig. 3.18. Be careful to keep the polarities the same on the loudspeakers, especially those in the same room. Check carefully for unintended shorts between different parts of the circuit or the metal of the case.

Before you try out the splitter, double check carefully for any mistakes. Then switch on the amplifier and try playing a tape or CD through the splitter with all the speakers in circuit, and also with just the main pair connected. If problems do occur, keep the amplifier volume low if you have to use it while you look for the cause. Here are some possibilities:

Switch has no effect: switch damaged by soldering, solder bridge across the contacts.

One speaker dead: poor connection in terminal block, short across loudspeaker, break in loudspeaker lead.

Resistors heat up and burn out: fault with amplifier (use a multimeter to check that its output is zero volts DC and AC with no input), high power amplifier and speakers being used (use higher power resistors).

TABLE 3.1

Parts List—Speaker splitter
Resistors R1,2,3,11,12,13   15Ω 1W carbon or carbon film
Miscellaneous SW1   2-pole, 2-way toggle switch, sub-minature
Box to suit (interior dimensions at least 25 × 70 × 100mm; prototype was a 3501 instrument case from Maplin Electronics); two terminal blocks, 7 way each; sleeving for resistor leads; coloured wide for internal connections; solder.

CHAPTER 4

The Silicon World

To understand semiconductors, we must look at the structure of solids, and in particular that of silicon. Other substances like germanium and gallium arsenide are used to make semiconductors, but the vast majority of semiconductor devices use silicon.

In Chapter 2, we found that atoms have a nucleus and layers of electrons in a cloud around the nucleus. They bond to each other by sharing electrons in their outermost layer. Different atoms have different numbers of electrons in the bonding layer, and this affects how they bond together. Silicon has 14 protons in its nucleus, so it will tend to have 14 electrons in the electron cloud surrounding the nucleus. Four of these are in the outermost layer, and in a crystal of silicon bonds are formed by neighbouring atoms sharing their bonding electrons (Fig. 4.1). In very pure silicon, all the electrons are fixed, so no current can flow, unlike metals, where the bonding electrons are free to move.

Suppose some atoms of phosphorous replace some silicon atoms in the crystal. Phosphorous atoms can fit in because they are similar in size to silicon, but they have five atoms in the outermost shell, so only four of these will be involved in the bonding. The fifth will be only loosely attached to the 'parent' atom–so loosely that the electron is effectively free to move around the whole crystal. The parent atom is said to be ionised with a net positive charge balanced by the negative charge on the electron elsewhere in the crystal. If there are a number of phosphorous atoms in the crystal, there will be an equal number of free electrons, which allows current to flow.

Silicon which contains suitable atoms with five bonding electrons, like phosphorous, is called N-type semiconductor, meaning that it has negative electrons available to carry the current through the semiconductor. The electrons themselves are called *negative carriers* of current. The phosphorous is known as N-type impurity; other atoms with 5 outer electrons can act as N-type impurity. Gallium, like phosphorous, is sufficiently similar to silicon for a gallium atom to replace a silicon atom in the crystal. However, gallium has only three electrons in its outer shell and there will be a 'missing' electron, a gap in the bonds. Although electrons from neighbouring atoms cannot break free, they can swap around between each other. The gap can move around the crystal as if it were a free

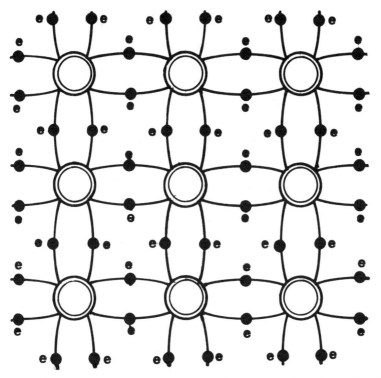

Fig. 4.1 A very simplified view of a silicon crystal. Each atom shares its own bonding electrons, one each with its four nearest neighbours, and in return has a share of one electronic from each neighbour.

electron with a positive charge; it is a *positive carrier* of current and is known as a *hole* (Fig. 4.2). The gallium atom left behind forms a negative ion.

Having a number of gallium ions in a silicon crystal creates an equal number of holes, so current can flow. Semiconductor with mobile holes is called P-type semiconductor, and gallium is called P-type impurity.

When P and N meet

When pieces of N-type and P-type semiconductor are joined, a P-N junction is formed; the characteristics of this junction are crucial to semiconductor electronics.

Electrons close to the junction on the N-type side and holes close to the junction on the P-type side combine and cancel each other out forming a region with no carriers called the *depletion zone* (Fig. 4.3). This zone extends into both the N-type and the P-type regions. Although there are no

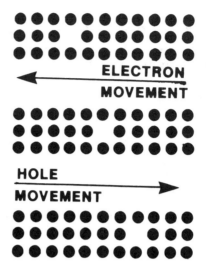

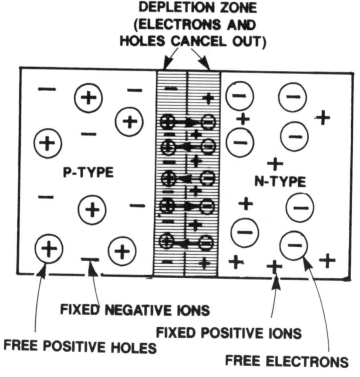

Fig. 4.2 A single hole moving along a semiconductor is actually lots of electrons moving one place in the opposite direction.

Fig. 4.3 Formation of a depletion zone between N-type and P-type semiconductor.

free carriers, the fixed ions remain, creating positive and negative charges on either side of the zone and repelling electrons and holes.

If a voltage is applied to the junction, with the P-type region made positive and the N-type region negative (Fig. 4.4), holes from one side and electrons from the other will be drawn towards the depletion zone. If the voltage is large enough, the charges in the depletion zone will be overcome and the holes and electrons will enter the zone, where they will meet and cancel each other out. At the connection between the N-type semiconductor and the voltage source, electrons will be entering the N-type semiconductor to replace the electrons crossing into the depletion zone. Similarly, at the connection between the P-type region and the supply, electrons will be leaving the P-type semiconductor creating holes to replace those crossing into the depletion zone. At every point, the flow of electrons (or holes in the opposite direction) is exactly the same.

Current can flow through the semiconductor junction in this direction provided the voltage is sufficient to overcome the charges in the depletion zone. The *threshold voltage* is the minimum voltage needed. It is a characteristic of the semiconductor material, and for silicon it is always

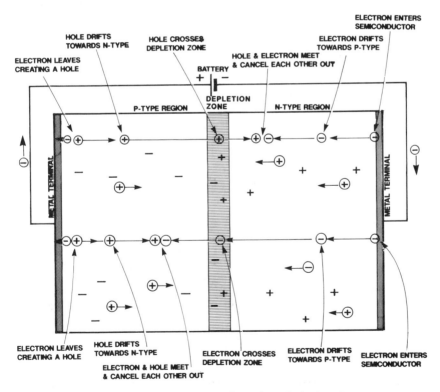

Fig. 4.4 The two means by which current flows through the junction.

around 0.6V. Above the threshold voltage, current flow increases very steeply. We can assume the voltage is 0.6V when small currents are being passed. The junction is said to be *forward biased* when a voltage is applied this way round.

Reversing the voltage

When a voltage is applied the opposite way round (Fig. 4.5), it pulls electrons away from the depletion zone on the N-type side, and holes away from the zone on the P-type side. As a result, no current flows and the depletion zone actually gets a little wider. A few electrons will be pulled into the voltage supply from the N-type region, and a few will be injected into the P-type region to cancel out some holes, but the net effect will be to charge up the semiconductors like the plates of a capacitor. In fact, a very small reverse or leakage current does flow, though for most purposes it is so small it can be ignored. It is caused by the occasional electron leaving the voltage supply on the P-type side and not meeting any holes until it crosses into the N-type region.

The overall result is that the P-N junction will *rectify*, it allows current to

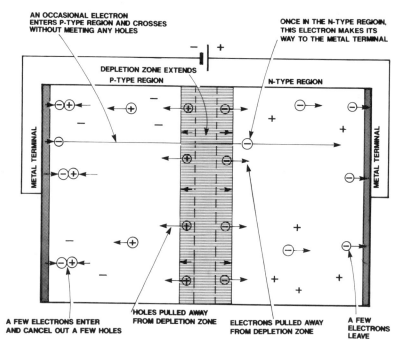

Fig. 4.5 Virtually no current can flow when the junction is connected in reverse to the battery.

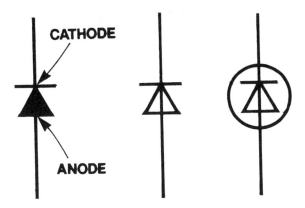

CATHODE

ANODE

Fig. 4.6 Different diode symbols. Conventional (ie positive to negative) current flows only in the direction of the arrow.

flow in one direction but not the other. Devices which do this are called *diodes*, and the common symbols for diodes are shown in Fig. 4.6.

Practical diodes

There are two common types of diode, *signal diodes* and *rectifier diodes*. Signal diodes are designed to have very low capacitance so that they will still work effectively at high frequencies. As already noted, the P-N junction resembles a small capacitor; signal diodes have a small junction area to minimise this effect, but as a result cannot carry large currents. Some signal diodes are also designed to have very low reverse currents.

Rectifier diodes are used to turn alternating voltages into direct voltages, i.e. to rectify them. They are designed to take large currents, so their junction areas are large, as are their capacitances, to allow lots of current to flow. However, since they operate at mains frequencies (50 or 60Hz), this is not important. A circuit using a transformer and a rectifier diode is shown in Fig. 4.7. The diode allows current to flow in only one direction, losing 0.6V; the output of this circuit is shown in Fig. 4.8.

This circuit has two disadvantages: less than half the available output from the transformer is used, and the actual direct voltage output varies considerably. Part of the solution is to use a bridge rectifier, Fig. 4.9, which has four rectifier diodes (this circuit is also called a full-wave rectifier). When A is positive in relation to B, diode D1 connects A to C and diode D3 connects B to D (because there are two diodes in the way a total of 1.2V is lost). When A is negative in relation to B, diode D2 connects B to C and diode D4 connects A to D. The resulting output is shown in Fig. 4.10: although an extra 0.6V has been lost, very nearly the whole of the waveform is used.

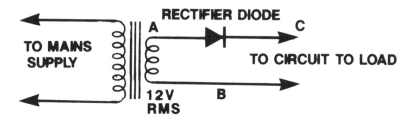

Fig. 4.7 A simple power supply circuit.

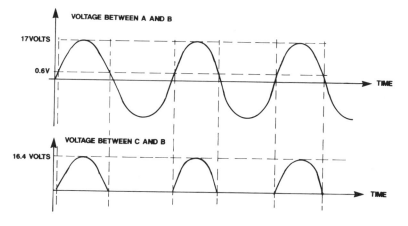

Fig. 4.8 The transformer secondary and output voltages of the circuit of Fig. 4.7.

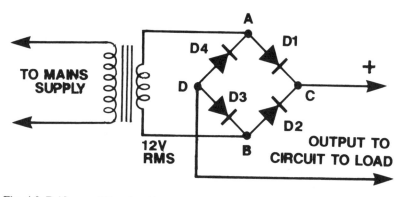

Fig. 4.9 Bridge rectifier circuit.

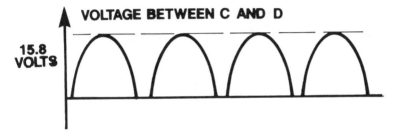

VOLTAGE BETWEEN C AND D

15.8 VOLTS

Fig. 4.10 Output from the bridge rectifier of Fig. 4.9.

Smoothing

The output from these circuits still is far from even, varying from zero to maximum and back again in a half-cycle. This is overcome by a smoothing capacitor connected across the output. Fig. 4.11 shows how a *smoothing capacitor*, C, is used with a bridge rectifier. For the sake of explanation, assume that the capacitor has no charge and the mains is applied to the transformer at a zero point in the cycle (Fig. 4.12). As A gets more positive in relation to B, current flows through diodes D1 and D3 charging up the capacitor. Once the maximum voltage is reached, the diodes cease to conduct because, due to the capacitor's charge, there is either no voltage or a negative voltage across them.

The voltage across the capacitor drops off slowly, depending on how much current is drawn by the load. At point Y', the capacitor voltage has dropped a little and is now low enough for diodes D2 and D4 to conduct (in this part of the cycle, B is positive and A is negative). The capacitor is charged up again to the maximum to point X', then discharges slowly until point Y is reached.

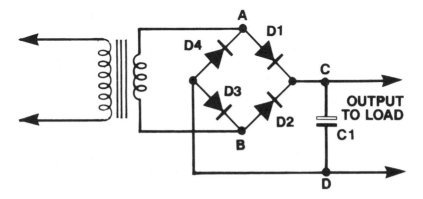

Fig. 4.11 How a smoothing capacitor is used with a bridge rectifier.

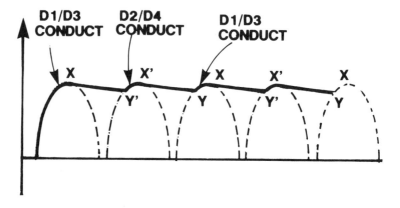

Fig. 4.12 The output of Fig. 4.11 still is not perfectly smooth, but has some ripple on it.

Ripple voltage

The amount by which the voltage across the capacitor changes between points X and Y' or X' and Y is called the *ripple voltage*. Suppose the maximum voltage at point X (or X') was 15.8V, if the voltage dropped to 14.8V by point Y (or Y'), the ripple voltage would be 1V (ripple voltages are usually quoted as peak-to-peak).

Estimating ripple voltages is quite easy and can save a great many problems, particularly when special voltage regulator ICs are used to make the voltage even steadier (these are discussed in Chapter 7). These require that the input voltage does not drop below a certain value even for a brief instant, so the ripple voltage has to be calculated. Normally the ripple voltage is a relatively small fraction of the total voltage, so the time the capacitor is being charged via the diodes (Y' to X') is very small, and the time the capacitor is discharging (X to Y') will be approximately half the cycle, 1/100th second in the UK or 1/120th second in the USA.

How far will the capacitor discharge in this time? The answer is given by the basic capacitor formula, $Q = CV$, where Q is the charge on the capacitor. A current of I amp flowing out of the capacitor will diminish this charge by I coulomb per second, or I/100 coulombs per half-cycle (I/120 USA). This makes V diminish by a corresponding amount, given by I/100C (I/120C USA).

An example will make this clearer. Suppose the capacitor in the circuit of Fig. 4.11 is 4,700μF, and the voltage of the transformer secondary is 12V RMS; if 1A is drawn from the supply, what will be the maximum and minimum output voltages? The maximum voltage will be the peak voltage of the secondary, $12 \times 1.42V = 17.04V$, less the voltage drop of two diodes; for rectifier circuits, it is better to assume 1V voltage drop per diode (as explained in the next section), which gives $17.04 - 2 = 15.04$V as the maximum voltage.

The ripple voltage is given by $V = I/100C$; I is 1A, C is 4,700uF or 0.0047F as must be used in the formula. The result is 2.13V (or, using $V = I/120C$ for the USA, 1.77V). This means that the minimum voltage from the supply will be $15.04 - 2.13 = 12.91$V (or $15.04 - 1.77 = 13.27$V USA). It so happens that this is much too low for a 12V regulator IC, so this calculation has already warned of a potential problem.

Ripple Current

Using a smoothing capacitor in a supply circuit causes the current to flow through the diodes in short bursts of very high current, rather than as a steady flow. Referring to Fig. 4.12, the diode conducts only for Y′ to X′ (or Y to X), and it must charge the capacitor sufficiently during this time for it to supply current for X′ to Y (or X to Y′). The lower the ripple voltage, the smaller the time from X′ to Y and the higher the current that must flow for this brief period. Typically, the burst of current will be of 10 times or more the average current flowing from the capacitor. This has consequences for every part of the circuit.

Firstly, the transformer must be able to withstand these pulses of very high current. In particular, the resistance of the windings must be very low. Suppose the resistance is an ohm, with an average current of 1A a ripple current of 10A will probably be flowing which will produce a voltage loss inside the transformer winding of 10V. The resistance clearly has to be much less than an ohm in this case. Secondly, the wiring between the transformer, diodes and capacitor has to use a substantial grade of wire. Using thin wire or inserting a fuse into this part of the circuit can lead to a large voltage drop. Thirdly, the diodes have to be able to take the very high ripple current; rectifier diodes are designed for this, and are normally specified according to the maximum average current they can take, so you do not have to worry about the ripple current. However, because of the high ripple current, the voltage drop across the diodes will be larger than the 0.6V normally assumed, typically around 1V.

Finally, the capacitor must also be able to take the ripple current. For low current power supplies this is not usually a problem, but if more than 250mA is drawn continuously, a capacitor with a high ripple voltage capability should be used.

Other diode types

Zener diodes behave as normal diodes in the forward conducting direction, but if a reverse voltage is applied above a certain voltage, a high current will flow. This critical voltage is called the *breakdown voltage* or *zener voltage*. If

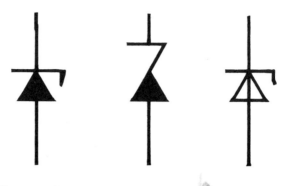

Fig. 4.13 Different symbols for zener diodes.

the applied voltage is reduced, the current falls away rapidly as the applied voltage goes below the breakdown voltage, reducing to zero slightly below the breakdown point. Zener diodes are very useful to provide a reference voltage, as shown in Fig. 4.14; in fact, this circuit can be used to provide a regulated supply if the current required is very low (a few milliamps maximum).

Zener diodes are available with a range of breakdown voltages from 2.7V up to 75V. Most of the values available fall into the E12 series, like resistor values.

Light emitting diodes (LEDs) emit light when they conduct in the forward direction (Figs. 4.15, 4.16). The forward voltage drop is around 2V rather than the usual 0.6V. The current required to give off enough light to be seen varies from a few milliamps to around 30mA. LEDs have the advantage over small filament bulbs in that they require less current and are more

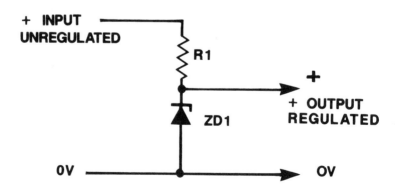

Fig. 4.14 How a zener diode can be used to provide a steady exact voltage from an unregulated supply such as Fig. 4.12.

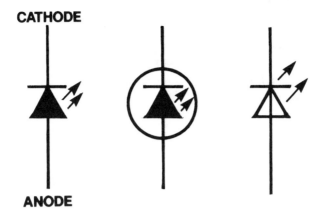

Fig. 4.15 Different LED symbols.

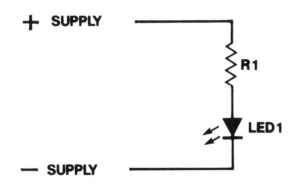

Fig. 4.16 How an LED is typically used.

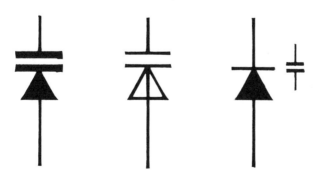

Fig. 4.17 Different symbols for varicap diodes.

robust. They are used extensively as small indicator lights and for numeric displays.

LEDs are widely available in four colours: red (the most common), orange, yellow, and green; infra-red LEDs are also available. LEDs usually have a resistor added in series with them to limit the current through them, or damage may occur.

Varicap diodes (also called varactor diodes) use the capacitance of a reverse-biased diode (Fig. 4.17). The width of the depletion zone varies according to the voltage applied across the diode, so the value of this capacitance changes. This property is enhanced in varicap diodes so that as the reverse voltage varies from 1V to 20V, the capacitance typically varies from 100pF to 10pF. Varicap diodes are used extensively in radio circuits.

CHAPTER 5

Silicon in Action: Transistors

Transistors are devices that allow a small current or voltage to control a larger current or voltage. They can be used as amplifiers, to make signals larger, or as switches, to allow a small current to control a much larger one. In this chapter we will mainly be concerned with the first use, but the second use is very important in digital electronics.

Two sorts of transistor exist: *bipolar transistors*, widely shortened to transistors, and *field effect transistors*, widely shortened to FETs. Bipolar transistors were in common use first; they are used both as discrete (single) components, each transistor with its own case and leads, and in integrated circuits, where many transistors and other components are ready-wired together in a single piece of silicon. The major use of FETs is in integrated circuits, though they can also be used in discrete form.

Bipolar transistor structure

Bipolar transistors are sandwiches of N-type and P-type semiconductor, forming two diode junctions back-to-back. An *NPN transistor* (Fig. 5.1) is two pieces of N-type with a piece of P-type between them. A *PNP transistor* (not shown) is two pieces of P-type with a piece of N-type in between. NPN transistors are slightly more common than PNPs, so we will study NPN transistor action. PNP transistors work in the same way, but with the roles of holes and electrons reversed.

The three layers of semiconductor are known respectively as the *emitter, base* and *collector*. Fig. 5.2 shows an NPN transistor connected to batteries which forward-bias the diode junction between the emitter and base and reverse-bias that between the base and collector. At the junction between the N-type emitter and the P-type base, electrons can move from the emitter through the depletion zone into the base. In a normal diode, they would meet and cancel out holes, but two factors prevent that from happening here: the base region is made very thin, typically a micron or less; and there are very few holes due to the very low level of P-type impurity put into the base during manufacture. The positive voltage on the collector region attracts the majority of electrons across the base and the depletion zone at the base-collector junction, and on into the collector itself.

Electrons can flow from the emitter, through the base and on into the

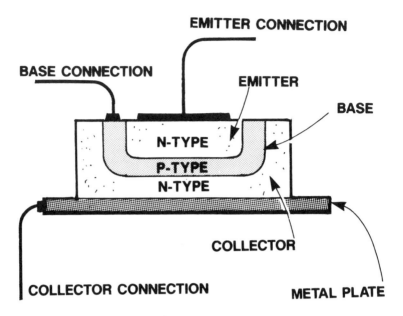

EMITTER CONNECTION

BASE CONNECTION

EMITTER

BASE

N-TYPE

P-TYPE

N-TYPE

COLLECTOR

COLLECTOR CONNECTION

METAL PLATE

Fig. 5.1 NPN transistor construction.

collector when the base-emitter junction is forward-biased. If this junction is not forward-biased, no electrons can flow whateverr the voltage on the collector may be. In terms of conventional current, current can flow into the collector when there is a sufficient forward-bias voltage on the base to make some current flow into it.

Current relations

The current which flows into the collector, symbol I_C, is normally a multiple of the base current, I_B, provided the battery between the emitter and collector can supply sufficient current. The ratio between I_C and I_B is called the current gain of the transistor, symbol β (Greek letter beta), h_{FE} or h_{fe}. h_{FE}, normally quoted in data books, is the total collector current divided by the total base current. h_{fe} relates small changes in collector current to small changes in base current.

In a perfect transistor, the current flowing into the collector would not be influenced by the voltage applied to the collector, as long as it was more than about 1V and less than the maximum the transistor was designed to stand. In practice, increasing the collector voltage does slightly increase the current. The emitter current equals the base current plus the collector current. The base current is usually so much smaller than the collector current that we can regard the emitter current as being equal to the

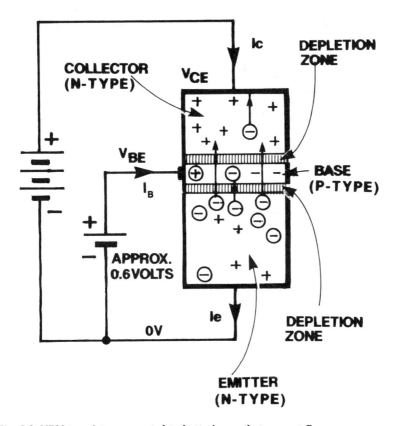

Fig. 5.2 NPN transistor connected to batteries so that current flows.

collector current. However, when the collector current is being limited externally, or the transistor has a very low current gain, the base current has to be taken into account. The voltage needed between the base and the emitter to make the transistor work at all is the diode threshold voltage, 0.6V.

Simple transistor amplifiers

One of the most widespread jobs for electronic circuits is to make a signal larger, and circuits that do this are called amplifiers. In theory, there are two sorts, *voltage amplifiers* which make a voltage signal larger, and *current amplifiers*, which amplify a current signal. In practice, this distinction is not so clearcut. For example, voltage amplifiers often have to provide extra current too.

Signals have some reference point, and this is usually the circuit's 'common' line. The signal is either a voltage difference with the common

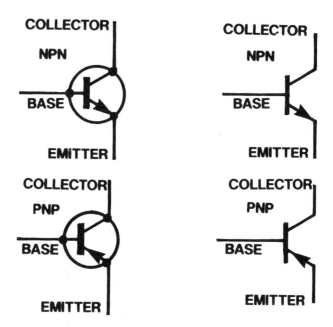

Fig. 5.3 The different symbols used for NPN and PNP bipolar transistors.

line or a current flowing to the common line. Also, the signal is very often added to a standing voltage or current which is needed to make the circuit work; the circuit will also add its own standing voltage or current to the output, which has to be taken into account later on.

Although the circuits here use only one or two transistors, they are the main building blocks used to make more complex circuits. In a single transistor amplifier, the input signal is applied to one of the transistor's terminals, and the output is taken from another terminal. The final terminal of the transistor is connected to the common line. The connection can be direct or through a capacitor, which is effectively a direct connection for alternating signals. The circuit is classified according to which terminal is connected to the common line. Only versions of these circuits using NPN transistors are shown, however, PNP versions are almost as common. The main difference with a PNP version is that all the currents and all the voltages flow in the opposite directions. Otherwise, the circuits are virtually identical.

Common emitter amplifier

The common emitter amplifier shown in Fig. 5.4a is a current amplifier, which has its emitter connected to the common line. Any current flowing

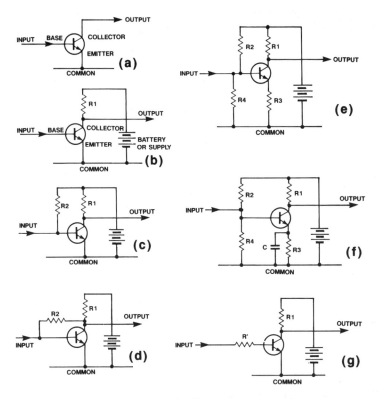

Fig. 5.4(a) Basic common emitter circuit; **(b)** common emitter with collector resistor added; **(c)** common emitter with base resistor to bias it on; **(d)** using negative feedback to control the operating point; **(e)** using an emitter resistor to control the operating point; **(f)** using a bypass capacitor to restore the gain; **(g)** using a resistor to convert voltage signals to current and increase input impedance.

into the base will cause a larger current (h_{FE} times the base current) to flow into the collector. As we have seen, the collector does not generate the current itself, but allows current from an external supply to flow. In Fig. 5.4b, collector current is provided by the battery and resistor R1. R1 also makes the collector voltage depend on the amount of current flowing. The transistor itself does not fix the collector voltage, the voltage just has to be large enough to sweep up the electrons crossing the base – a fraction of a volt is enough. However, R1 has the voltage across it and current through it (the collector current) linked by Ohm's Law, so the voltage has to be the resistance times the current:

$$V_{R1} = R1 \times I_C$$

The voltage at the top of R1 is the battery voltage which is fixed, so the collector voltage is the resistor voltage subtracted from the battery voltage:

$$V_C = V_B - V_R = V_B - R1 \times I_C$$

The higher the base current, the higher the collector current, but the lower the collector voltage.

Getting a circuit like this to work properly depends on carefully selecting the *operating point* – the combination of standing currents (or voltages) which are added to the signal to make sure there is enough current flowing for the circuits to work at all. The standing currents in the base and collector are the main problem. They have to be large enough for the negative peak of the signal not to cut off the transistor (i.e. cancel out the standing base current, so that no current flows at all). However, they should not be so large that the positive peak of the signal causes *saturation*. Saturation means that the maximum possible collector current that can flow is already flowing. The upper limit is the battery voltage divided by the resistor's value.

The simplest way to set the operating point is shown in Fig. 5.4c. R2 supplies a standing current to the transistor's base, which is added to the signal current. This circuit gives a large signal gain – a very large output for a small input. One problem is that current gains of transistors vary a great deal, even when they are supposed to be the same. The right value of R2 for one BC184L transistor will probably be wrong for the next BC184L taken from the same box (and wrong for the original transistor when its temperature is higher or lower).

Fig. 5.4d shows an alternative arrangement which gets over this problem. If the collector voltage is too high, the voltage across R2 increases, but this increases the current into the base and brings down the collector voltage. This is *negative feedback* – a portion of the output is fed back to the input as a way of correcting errors. But negative feedback also affects what happens to the input signal, reducing the gain but making the amplification much more accurate.

Another circuit which gets round the problem of transistor variation is shown in Fig. 5.4e. Extra resistors R2 and R4 form a potential divider which supplies a fixed proportion of the battery voltage to the transistor base. (If the signal already has a standing voltage added to it, say by another amplifier, R2 and R4 may not be needed.) The voltage at the transistor's base has to be the base voltage less 0.6V; all the current has to flow through R3 and because the emitter voltage is fixed, the current is also. Almost exactly the same current (except for the very small base current) flows through R1. Any variations in base voltage cause identical variations in the emitter voltage, and, when R1 is larger than R3, amplified variations of the collector voltage in the opposite direction. This circuit is actually a voltage amplifier, since both input and output are voltages, but it inverts the output.

All the circuits so far can be used with AC or DC signals, although for DC signals standing currents and voltages have to be added or subtracted. The circuit of Fig. 5.4f is suitable for AC signals only, but it offers high signal gain with an easily set operating point. For DC, the circuit's action is the same as Fig. 5.4e, so the operating point can be set easily, but for alternating signals, capacitor C connects the transistor's emitter directly to the common line, making its action the same as Fig. 5.4b or Fig. 5.4c. The value reactance of C must be small – a few ohms or less – at the lowest frequency we wish to amplify.

Bypass capacitors can be used to pass alternating signals around while avoiding standing voltages. This does not work for DC signals, so we use multi-transistor circuits where all the standing voltages cancel each other out.

In all the circuits mentioned except Fig. 5.4e, the input signal has been a current not a voltage. In practice, the distinction is not so clearcut because there will always be some resistance in the signal path – either the internal resistance of the preceding stage or in the transistor base's internal resistance. The circuit in Fig. 5.4g uses a resistor R5 to convert a voltage signal into a current. When the emitter is connected to common, the base seems to input signals like a diode connected to common. Sometimes this can cause excessive current to flow, and this is stopped by R5. R5 increases the input impedance of the circuit.

Common collector amplifier

It may not appear that the collector of the transistor in Fig. 5.5 is 'common', i.e. connected to the common line. However, the battery holds it at a constant voltage with respect to the common line.

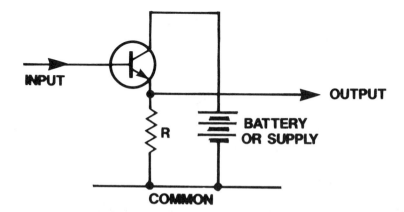

Fig. 5.5 The common collector circuit, very often called the emitter follower.

This circuit is almost always called an *emitter follower*. Because the base and emitter of the transistor are always 0.6V apart, the emitter voltage 'follows' the input base voltage. This circuit is a voltage amplifier but with a gain of 1. It takes only a very small current from its input but can supply a much larger output current (the input current multiplied by the transistor's current gain). The circuit has a high input impedance and a low output impedance, and this gives it, and circuits like it, a very wide range of applications.

Common base amplifier

This amplifier (Fig. 5.6) is rarely used; it performs the opposite function to the common base circuit, offering a very low input impedance and a high output impedance. Its main use is in radio circuits, but the Superior Continuity Tester project at the end of this chapter also uses it.

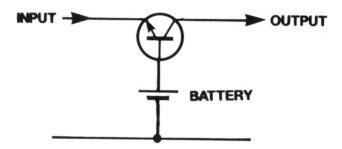

Fig. 5.6 Basic circuit of the rarely-used common base amplifier.

The transistor's base voltage is held constant, making the emitter voltage 0.6V lower and effectively constant whatever current flows through it. However, putting a resistor in series with the collector will produce variations in output voltage.

Circuits for two transistors

The two most common circuits which combine two transistors are the long-tailed pair (Fig. 5.7) and Darlington pair (Fig. 5.8). A long-tailed pair is a voltage amplifier. In its simplest form, it has two identical transistors (Q1 and Q2) with identical, high current gains, and two identical collector resistors R1 and R2. However, the emitter resistor R3 is shared. Standing voltages are applied to the two inputs V_{IN1} and V_{IN2} so that V_T, the voltage across R3, is about half the supply voltage, V_{CC}. The standing output voltages at outputs V_{OUT1} and V_{OUT2} are about halfway between V_T and V_{CC}.

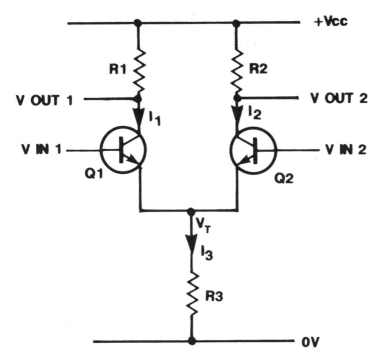

Fig. 5.7 Long-tailed pair circuit, used very frequently as part of an integrated circuit.

The emitter resistor R3 is the 'long tail' because it has a relatively high voltage across it (15V for a 30V supply); in comparison, a typical input signal would be several tenths of a volt. Because signals change V_T only very slightly, the total current flowing through the two transistors must remain the same. However, the signals do make a considerable difference to the currents through the individual transistors, and divert current from one transistor to the other. Suppose a positive signal at V_{IN1} increases the current through Q1. This reduces the voltage of V_{OUT1} but increases the voltage at V_{OUT2}. The signal at V_{OUT1} is inverted with respect to V_{IN1} but V_{OUT2} is non-inverted. The circuit is symmetric so that a signal V_{OUT2} is inverted with respect to V_{IN2}, but V_{OUT1} is not inverted. The circuit amplifies the *difference* between inputs V_{IN1} and V_{IN2}. If both inputs increased by the same amount, the changes cancel each other. This circuit is therefore said to have *differential inputs*.

Currently long-tailed pairs are rarely made using separate components, but they are very widely used in integrated circuits.

In the Darlington pair in Fig. 5.8, the emitter current from Q1 flows straight into the base of the second transistor effectively making the two together a 'super' transistor with a current gain equal to the two individual

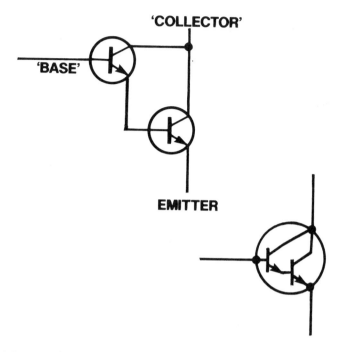

Fig. 5.8 Darlington pair 'super' transistor.

gains multiplied together. The base-emitter voltage needed to bias this 'transistor' on is now 1.2V, twice the usual. This arrangement is used often with high-power circuits, because power transistors usually have relatively low current gains. Using a small high-gain transistor as Q1 gives a very useful increase in gain – so useful that Darlington pairs are commonly available housed in a single case. Both NPN and PNP versions are used widely.

Field effect transistors

FETs work in a different way from bipolar transistors. We saw in the previous chapter that as greater and lesser reverse voltages are applied across a diode junction, the depletion zone will widen or narrow, and this effect was used to make varicap diodes. FETs use this effect in a different way. An FET called an N-channel junction FET (J-FET) is shown in Fig. 5.9. The N-type semiconductor in the diode's cathode has been stretched to form a channel which runs between two contacts, the *source* and the *drain*. The anode of P-type semiconductor runs above and below the channel, and between them is a junction with a depletion zone.

Usually the source is negative with respect to the drain, so carriers

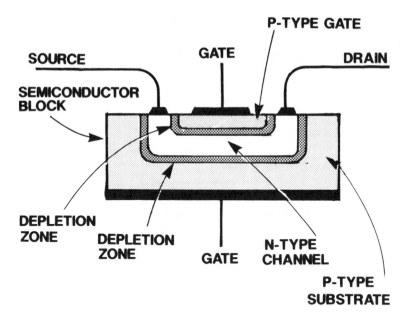

Fig. 5.9 N-channel junction FET.

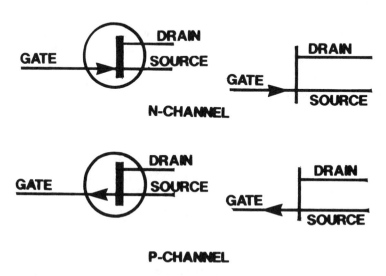

Fig. 5.10 Different JFET symbols.

(electrons in N-type semiconductor) flow through the channel from the source to the drain (in simpler JFETs, the source and drain are interchangeable). The gate is made negative with respect to the whole channel (including the source and drain), so that no current can flow between the gate and the channel. (Actually, there is a very small leakage current, as there is with any diode junction, but this is so small that most of the time we do not have to worry about it.)

Making the voltage more negative increases the width of the depletion zone, and it will *pinch off* the channel completely, so no current can flow when the *pinch-off voltage* is reached. Below this voltage, the channel width varies according to the gate voltage, so the size of current flowing is linked to the gate voltage.

The FET has two modes of behaviour. With a low voltage between the source and drain, the FET is in its *linear* region and the channel acts like a resistor whose value is controlled by the gate voltage (a voltage-controlled resistor). When the drain-to-source voltage exceeds a few volts (exactly when depends on FET type), the FET is in its *saturated* region: the current flow is governed by gate voltage alone – as long as the drain to source voltage stays above a minimum level, increasing it further does not increase the current flow. The FET's *transconductance* is the amount the current increases for an increase (making it *less* negative) in gate voltage. Transconductance is measured in amps per volt or *mhos* (ohms spelt backwards with the 's' moved to the end); actually micro-mhos (μmhos or umhos) are more common. The correct SI unit for amps per volt is Siemens (S) but mhos are more commonly used. FETs are used mostly in the saturated mode, though they are used occasionally as voltage-controlled resistors in the linear mode.

For N-channel JFETs, the gate must always be kept negative with respect to both the source and drain, or the internal diode will conduct and the FET action will be ruined. P-channel JFETs do exist, and their gates must always be kept positive; their drains are normally negative with respect to their sources.

MOSFETs

MOSFET stands for Metal Oxide Semiconductor Field Effect Transistor. The gate in a MOSFET is a thin strip of metal insulated from the semiconductor by a very thin layer of silicon oxide (Fig. 5.11). They are sometimes known as IGFETs (Insulated Gate FETs). We will look at the action of an N-channel MOSFET in detail, although P-channel MOSFETs are just as common.

Because the gate can be negative or positive with respect to the channel, two MOSFET types are possible. *Depletion mode MOSFETs* operate in exactly the same way as J-FETs, but have very much lower leakage current

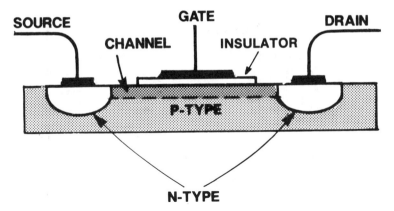

Fig. 5.11 N-channel enhancement mode MOSFET.

from the gate. *Enhancement mode MOSFETs* actually have only a depletion zone between the N-type source and drain areas until the channel is induced by making the gate positive with respect to the source. As usual, the width of the channel depends on the gate voltage. MOSFETs are very vulnerable to static charges building up on their gates and destroying the oxide layer, because their gates are so well insulated. Special precautions have to be taken in handling them to prevent this happening.

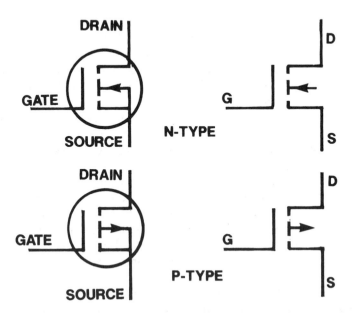

Fig. 5.12 The symbols for enhancement mode MOSFETs.

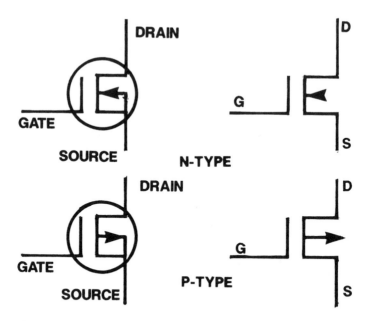

Fig. 5.13 Depletion mode MOSFET symbols. Often, though, circuit diagrams do not distinguish between the two types of MOSFET and this symbol is used for both.

Using FETs

Many of the circuits designed for bipolar transistors can be modified for JFETs and MOSFETs. The biasing voltages may have to be changed but the basic circuit is the same. For instance, Fig. 5.14 shows a common source circuit using an N-channel JFET (the same circuit could be used with a depletion-mode N-channel MOSFET). Note that R2 keeps the gate voltage negative with respect to the source. If an enhancement mode MOSFET was used, the gate would have to be biased positive with respect to the source with a potential divider (like the base in Fig. 5.6f) and R3 and C could be left out.

The FET circuit has the advantage over the transistor circuit of a very high input impedance (several megohms is not unusual, compared to a few kilohms for the transistor version). However, FET characteristics are even less consistent than those of transistors, so setting the operating point is more difficult. The gain of this simple circuit is also lower – about 10 or 20 as opposed to 100 or more for the transistor circuit. An N-channel FET can be used to substitute directly for the transistor in the emitter follower circuit of Fig. 5.8, making a source follower. The voltage which the source assumes will be determined by the type of FET used (it will be above the gate voltage for JFETs and depletion mode MOSFETs, but below it for

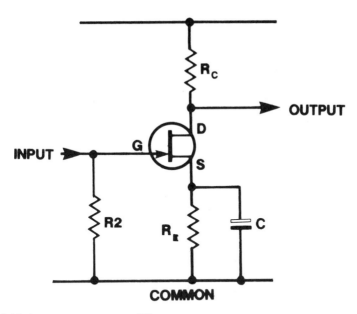

Fig. 5.14 A common-source amplifier.

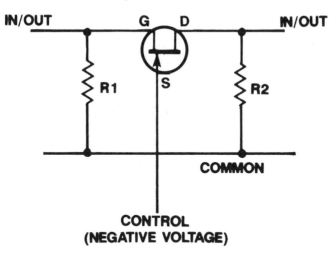

Fig. 5.15 Using a JFET as a voltage controlled resistor.

enhancement mode MOSFETs). Again, the exact voltage will vary a great deal from FET to FET.

Fig. 5.15 shows a FET being used as a voltage controlled resistor. The two IN/OUT terminals, the drain and source, must have no overall DC voltage between them, except that caused by the signal (and this must not exceed 1V) so that the FET operates in its linear region. The gate voltage must be negative with respect to both; the more negative it is, the higher the resistance between source and drain.

PROJECT: Superior continuity tester

Most continuity testers cannot distinguish between a good connection (an ohm or under) and a bad one (several hundred ohms or more). The superior continuity tester can.

The basic circuit of the continuity tester (Fig 5.16) is on a common base transistor circuit. The current through the collector resistor, R_C, is controlled by the emitter resistor, R_E. If the voltage of the base is set at very slightly above 0.6V, changing R_E from zero to an ohm or less will make a very big difference to the current flowing through the collector. The circuit is very sensitive to low values of R_E, when 100mA or more can flow, so Q1 has to be a power transistor.

In Fig. 5.17 R_E is replaced by two test connections. Two diodes D1 and 2, potentiometer RV1 and resistor R1 are used to provide an adjustable base voltage of 0.6 to 1.2V. When RV1 is correctly adjusted, a change in resistance between the test terminals from zero to an ohm will make an easily detectable difference to the current flowing through Q1 and the voltage across R2.

In the final circuit of Fig. 5.18, ZD1, R3 and Q2 are added to detect when the collector voltage falls below a certain level and make the sounder work when this happens. PNP transistor Q2 is used in a common emitter circuit, and zener diode ZD1 stops Q2 conducting until the collector of Q1 goes below 5.7V, when the current through R2 is 30mA. Diode D3 has also been added to prevent the emitter voltage going above the base voltage, and an

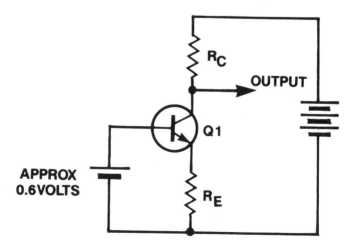

Fig. 5.16 Basic continuity tester circuit.

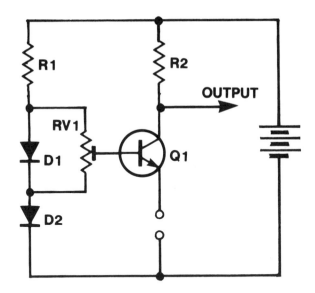

Fig. 5.17 Making the base voltage adjustable.

LED and resistor are added to provide an indication of when the circuit is turned on. A reference resistor, R5, is included for checking operation.

Buying
The parts needed for the project are described in the parts list (Table 5.1). Some sources for these are given in Chapter 9, including details of how to obtain the specially designed printed circuit board (PCB). Although the project can be built without the PCB, it is not recommended since it is very easy to make mistakes in the wiring.

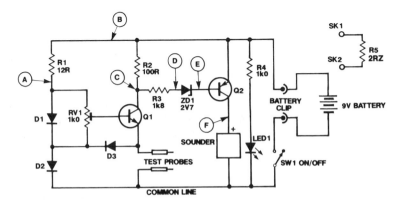

Fig. 5.18 Final circuit with test points marked.

TABLE 5.1

Parts List — Superior continuity tester
Resistors (all at least $\frac{1}{4}$W, 5% or better)
R1   12k
R2   100R, $\frac{1}{2}$W
R3   1k8
R4   1k0
R5   2R2
RV1   1k0 preset potentiometer, sub-miniature, horizontal mounting
Semiconductors
Q1   BD139, MJE340 or 2SC1162
Q2   2N3702, 2N3906 or BC214L
D1,2,3 1N914 or 1N4148
LED1   red LED, $\frac{1}{4}$ inch (5mm) type, with mounting clip
Miscellaneous
SW1   SPST (or SPDT) switch, sub-miniature toggle
6V to 9V sounder, maximum current 50mA or less; PP3 battery and clip; PCB; PCB pins; test probes; extra-flexible wire (1m red, 1m black); test sockets for R5; box; glue; solder; wire.

All the resistors specified have values in the widely available E12 range. A tolerance of 5% or better should be used (eg 2% or 1%, though these will make no improvement to the circuit's action), and the resistors should able to at least dissipate $\frac{1}{4}$W, so 0.4W types could be substituted but not 1/8W. R2 should be rated at least $\frac{1}{2}$W. The potentiometer can be any sub-miniature type of the right value, but a miniature type will be too large. Alternatively, a full-size rotary pot could be mounted separately and wires used to connect it to the PCB. Alternative types are given for the transistors and diodes.

The sounder used in the prototype was a miniature solid state buzzer, but any buzzer which operates on 6 to 9V and draws 50mA or less will be acceptable (piezo-buzzers are not suitable unless they have internal driving circuitry). Sounders which can work over a larger voltage range can be used; those which work over a smaller range cannot. The case used in the prototype had a special compartment for the battery, and is described in catalogues variously as a 'small remote control box' or 'project box with battery compartment'.

Assembly
Start by mounting the PCB pins in the correct position and solder them – these are for attaching leads. You can get a special tool to insert the

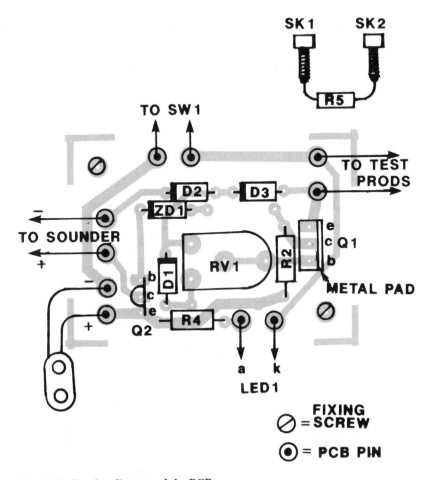

Fig. 5.19 Overlay diagram of the PCB.

pins, but I find a standard pair of pliers just as easy. The protruding ring on the pins should be on the component side (the side without copper tracks); once soldered, this prevents the copper track being pulled off the board when leads are tugged.

Position the resistors with their leads poking through to the copper side. Solder them, then cut off the excess lead close to the soldered joint. Position and solder the diodes (including the zener) taking great care to get the bands on their bodies at the ends marked in the overlay diagram. Solder the transistors into position, leaving about 5 to 10mm of lead between the transistor bodies and the PCB; the correct orientation is shown in the overlay diagram (note that they are viewed here from the component side of the PCB). Complete the circuit by wiring the battery clip and sounder

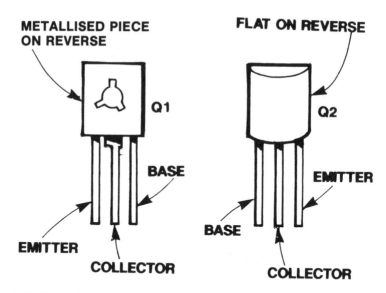

METALLISED PIECE ON REVERSE

Q1

BASE

EMITTER

COLLECTOR

FLAT ON REVERSE

Q2

EMITTER

BASE

COLLECTOR

Fig. 5.20 Transistor connections for the types given in the Parts List.

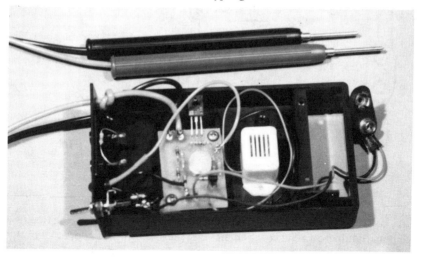

Fig. 5.21 The inside of the Continuity Tester.

leads to the pins on circuit board as shown. The positive and negative leads must be distinguished carefully. Solder test leads to the board.

Now stop and check all the circuit very thoroughly. Look for bad joints and solder bridges. Look out for silly mistakes – usually the hardest to spot! Then cover up the overlay and circuit diagrams, and work out with a pen and paper what the circuit you have actually made is. If possible, get a friend to check the circuit for you. Once you are satisfied that the circuit is

Fig. 5.22 The finished project.

exactly what it is supposed to be, connect the battery and try it out. Adjust RV1, if necessary, to make the sounder come on. If the sounder will not come on, disconnect the battery and start looking for the reason. Very often, the problem is very obvious – so obvious that you overlooked it first time round.

Fault tracing
If you cannot find the fault by looking, attach the negative multimeter lead to the negative supply lead and reconnect the battery. Switch the meter to a 10 volts DC range and probe around the circuit, comparing your readings with the value given in Table 5.2; all were taken with RV1 fully clockwise, so adjust your RV1 to this position. R2 will get quite hot, and a lot of current should be drawn from the battery while the test probes are together, so I recommend switching off between readings.

If the voltage at A is much lower than the value in the Table, there is probably a short somewhere, possibly across D1 or D2. If the voltage is over 1V (or over 1.4V with the prods apart) there is probably a bad connection. If the voltage at B is much below 7V, the battery could be old, there could be a poor connection between the battery and the circuit, or there could be a short circuit somewhere. This circuit requires a fairly new battery, which should read 9V or more with nothing connected (a nearly dead battery will have 7V or less).

TABLE 5.2

Test voltages for fault finding		
Point	Probes together	Probes apart
A	0.75V	1.2V
B	8V	9V
C	0V	9V
D	4.7V	9V
E	7.3V	8.4V
F	7.6V	0V

If the voltages at A and B are OK, check C: if it is above 1V, there is a bad connection to the transistor or the test prods, or transistor Q1 may have been damaged during soldering. If C is below 1V, check D; if it is the same voltage as C, there is a bad connection between D, diode ZD1 and transistor Q2, or Q2 has been damaged. If D is about the same voltage as B, there is a bad connection at R3. E should be 0.6 to 0.7V lower than B: if it is the same voltage as B, Q2 is faulty, there is bad connection to ZD1 or Q2 itself, or ZD1 has been damaged; if it is 0V, Q2 has been damaged or has a bad connection. Finally, if the voltage at E is OK but that at F is zero, there is a fault in Q2 or the sounder is shorted. If F is the correct voltage, but no sound comes from the sounder, it may be connected round the wrong way.

This should cover most eventualities, but not all. The gremlins who put faults into electronic projects have a never-ending supply of ideas!

The way the project is installed in its case is shown in Fig. 5.21. Epoxy resin (Araldite) can be used to glue the sounder in position, and posts to mount the PCB can be improvised from plastic shelf-mounting studs, also glued in position.

Setting up

R4, mounted between the two sockets on the top panel of the case, is used to set up the meter. With a test probe on each socket, adjust RV1 just to the point where, if it was turned any more, the sounder would come on. Then put the test probes together and check that the sounder comes on, indicating a good connection. This calibration should be repeated every time the tester is used, and also every 15 minutes or so, during prolonged use. To make this possible a hole, just large enough to get a narrow-bladed screwdriver through, should be drilled in the case above RV1.

It is possible to set up the prototype with an R4 of 1Ω, but results could then be inconsistent. However, you may find it useful to have a 1Ω resistor to see if you can distinguish between this resistance and no resistance at all.

CHAPTER 6

Silicon in Action: Integrated Circuits

Integrated circuits (ICs) are ready-made 'building blocks' containing collections of transistors and associated resistors and capacitors. These components together make up all or part of the circuit needed to do some specific job, for example they might be a radio or a computer memory.

As with single transistors, at the heart of each IC is a small piece of silicon – the chip. But on the surface of this chip, a few millimetres across, up to a million transistors may have been manufactured and linked together by narrow metal tracks. Resistors are made from areas of P-type or N-type semiconductor, and very small capacitors are formed by overlapping two areas of track separated by oxide. The dimensions of the transistors and other components – and the widths of the tracks linking them – are measured in microns (millionths of a metre), and the latest ICs use dimensions of under a micron.

Types of IC

We can split ICs into two broad types: *analogue* (also called *linear*) and *digital*. Analogue ICs are designed so that a small change in the input produces a small change at the output, although the change will be altered in some way, perhaps amplified (the job of the IC is normally to alter the signal). Digital circuits are very different; input and output voltages represent binary numbers and are either 'on' or 'off' ('high' or 'low'). They are discussed in Chapter 8.

Both analogue and digital ICs can be *special purpose,* made to do one very specific job (for example, a car stereo amplifier), or *general purpose,* made to perform an electronic function which is used in different circuits. Much of this chapter will be taken up by op-amps which are general purpose ICs, and one of the commonest building blocks in analogue electronics.

Op-amps

Op-amps are voltage amplifiers. The earliest were made from individual transistors and other components, but today's are nearly all ICs. The symbol for an op-amp is shown in Fig. 6.1. There are two inputs called the *inverting input* and the *non-inverting input;* however there is a single output. There are two power supply connections. The op-amp takes the *difference* between the two input voltages and its output is a greatly amplified version of it relative to the common line. If the non-inverting input is higher than the inverting input, the output will be positive; if the non-inverting input is the lower, the output will be negative. Ideally, the output will be zero whenever the two inputs are actually the same whatever voltage they are.

The op-amp has a very high voltage gain, called the *open-loop gain.* Gains of 100,000 are typical, and most of the time we can think of the gain as so large that the exact figure is unimportant. However, the output voltage cannot be higher than the positive supply voltage or lower than the negative one. So with supplies of ±12V, the output cannot be +100V or even –12.5V. The very high gain means that a difference of a few microvolts between the two inputs makes the output voltage go to one or other of the supply rails, which is known as *saturating.* In fact, slight differences between the inverting and non-inverting input circuitry mean that the output will often be saturated even when the inputs are the same. We use *negative feedback* to get a more useful output.

Op-amps usually work from two equal positive and negative supply voltages (dual supplies) with a common line which is not directly connected to the op-amp. The minimum supply voltage is typically +5V on the positive and –5V on the negative (usually written ±5V); the maximum is usually ±15V. Some can work from as little as ±1.25V, and others have higher maximum voltages. Most can work from a single power supply

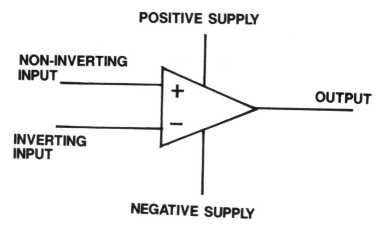

Fig. 6.1 Circuit symbol and connections to an op-amp; note that the positions of the two inputs can be swapped, depending on which is most convenient—look for the + and – signs to distinguish them.

voltage, and some are specifically designed to do so. To make circuit diagrams easier to follow, op-amp supply connections are often not shown, but they must always be present for the op-amp to work.

Op-amp as a buffer amplifier

Fig. 6.2 shows the simplest op-amp negative feedback circuit. The input is applied to the non-inverting input and the output is fed back to the inverting input. All input and output voltages are with respect to the common line.

Suppose that the input voltage to the circuit is lower than the output voltage. A difference between the two inputs, with the non-inverting input lower than the inverting input, would normally cause the output to saturate at the negative supply line. The output voltage therefore begins to move towards the negative supply voltage, but before it gets there it reaches the input voltage. At this point, there will no longer be any difference between the op-amp's input voltages and the output voltage will stop changing. (The output voltage will actually stop a few microvolts away from the input voltage, just enough to give the difference in inputs required for the output voltage to be anything other than zero. Vital though this difference is to make the op-amp work, it can usually be ignored in calculations.) If the input voltage changes, the output voltage will change with it, following it.

The op-amp is comparing the non-inverting input voltage to the output voltage fed back to the inverting input, and correcting itself to make the two equal. This circuit, called a *voltage follower,* is the equivalent of an emitter follower but offers even higher input impedance and lower output impedance. There is no standing voltage difference between the input and output (actually, there is but only a few microvolts), and both can be positive or negative, and can go to within a couple of volts of either power supply. Perhaps you can already see why op-amps are so useful!

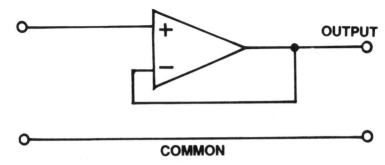

Fig. 6.2 Connecting the output to the inverting input.

Changing the gain

In Fig. 6.3, only part of the output is fed back to the input; resistors R1 and R2 act as a voltage divider, feeding back just a proportion R1/(R1+R2) of the output to the input (1/11th for the values shown). The op-amp will adjust itself so that its two inputs are equal, but to do this the output has to be 11 times the input. The circuit therefore has a gain of 11. In general, the gain of a circuit like this is (R1+R2)/R1. However, remember that the op-amp's output cannot go beyond the power supply voltages, regardless of input voltage and circuit gain. If a signal tries to take the output beyond the limits of the power supply, it will *clip*. The top and/or bottom of the output signal will be limited to the supply line voltages, but the rest of the signal will be the same as usual.

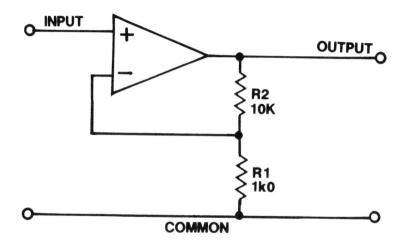

Fig. 6.3 Feeding back just part of the output.

Inverting amplifier

The circuit above is non-inverting – a positive input will give a positive output. Op-amps can also give inverted outputs, and Fig. 6.4 shows the basic circuit. As before, the feedback is to the inverting input of the op-amp (via R2) so that it tries to make its input voltages equal; as the non-inverting input is tied to the common line, the inverting input must be at 0V when the op-amp succeeds. However, any voltage applied to the input on R1 will tend to move the op-amp's inverting input voltage away from the common voltage, so the output voltage 'pulls' against this via R2 to make the inverting input zero. The higher the value of R2 and the lower the value of R1, the more the op-amp's output has to 'pull' to make the inverting input zero, so the greater the circuit's gain.

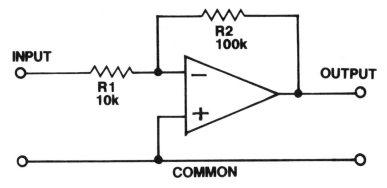

Fig. 6.4 Basic inverting amplifier.

For a perfect op-amp, there will be no current flowing into the inverting input because the input impedance is so high. The current flowing in R1 must therefore flow through R2. The current through R1 is $(V_{IN}-0)/R1$ and that through R2 is $(0-V_{OUT})/R2$ (one or other of V_{IN} or V_{OUT} must be negative). As these are the same:

$$\frac{V_{IN}}{R1} = -\frac{V_{OUT}}{R2}$$

$$V_{OUT} = -V_{IN} \times \frac{R2}{R1}$$

So the circuit has a gain of $-R2/R1$. The negative sign indicates that the output is inverted with respect to the input. For the values shown in Fig.

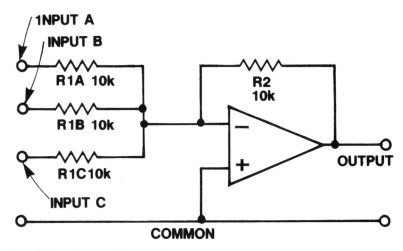

Fig. 6.5 Inverting amplifier used as a mixer.

6.4, the gain will be −100/10 = −10; an input voltage will produce an output voltage ten times the size but inverted (always remembering that the output voltage cannot be greater than the supply voltages).

The inverting input of the op-amp in this circuit is said to be a *virtual earth.* Although it is not actually connected to common, the op-amp always acts to make it exactly equal to common. Fig. 6.5 makes use of this feature to mix three inputs together. The advantage of this circuit is that the inputs are isolated from each other by the virtual earth point. Traces of one input do not appear at the other inputs, so input A could be connected to the input of another amplifier as well without the other inputs coming through. This circuit is the basis for mixers used in the vast majority of studio mixing desks, public address systems (PAs) and sound systems.

Combined amplifier

The circuit of Fig. 6.6 is a *differential amplifier,* one which amplifies the difference between the two inputs like the op-amp itself, but the gain can be set to any required value by the resistor values. Normally we make the gains from the two inputs exactly equal but opposite. To see this, suppose input 2 is attached to the common line, the circuit is then the same as Fig. 6.5 with two extra resistors which make no effective difference. The gain for signals to input 1 is therefore −R2/R1. If input 1 is earthed, the circuit becomes the same as Fig. 6.3 but with a voltage divider, R3 and R4, added on the non-inverting input. The gain would be (R2+R1)/R1 without this, but the voltage divider reduces the input voltage by R4/(R3+R4), making the overall gain R4×(R2+R1)/R1×(R3+R4). If we make R3=R1 and R4=R2, this simplifies to R2/R1. The gain for signals to input 2 is the same as the gain from input 1 but positive.

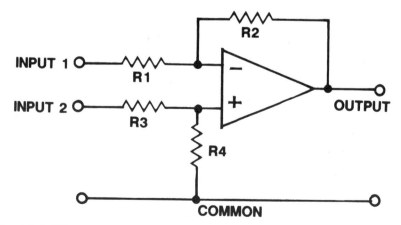

Fig. 6.6 A difference amplifier, combining non-inverting and inverting circuits.

Frequency selective feedback

There is no reason why feedback circuits should use only resistors. Using a capacitor with the feedback resistor, we can make the gain change according to signal frequency, something which is very useful in audio and other circuits. In Fig. 6.7, C1 is in series with R2b, and these two together are in parallel with R2a. The gain of the circuit will vary as shown in Fig. 6.8.

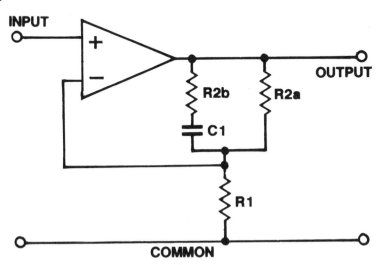

Fig. 6.7 Non-inverting amplifier with frequency-dependent gain.

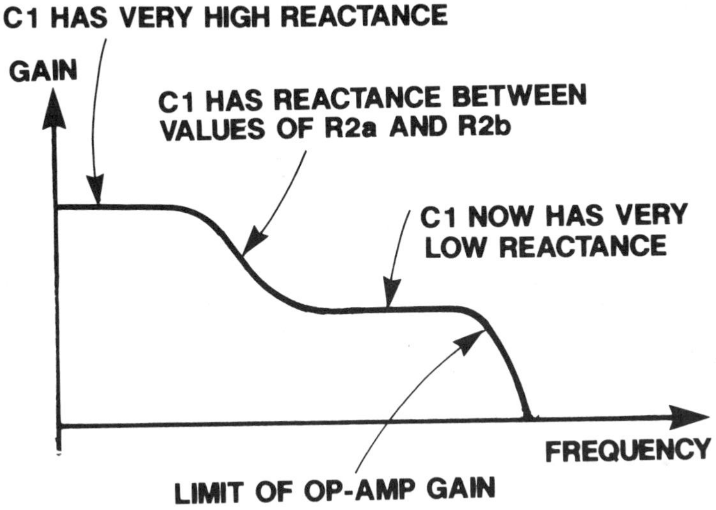

Fig. 6.8 The gain of the circuit of Fig. 6.9 at frequency changes.

For low frequency signals, the reactance of capacitor C1 is so large that resistor R2b is effectively disconnected, and the gain of the circuit will be (R1+R2a)/R1. For high frequencies, the reactance of the capacitor is so small that R2a and R2b are connected in parallel. If R2 is their combined resistance when in parallel, the gain is (R1+R2)/R1.

For intermediary frequencies, the reactance of C1 will be about the same size as the resistances, so it is in series with R2b and these are in parallel with R2a. Because it is a reactance, doing the calculation involves using phasor diagrams. However, it should be obvious that as frequency increases, the reactance decreases, so the combined impedance of C1, R2a and R2b decreases. The gain decreases with increasing frequency.

We could obtain the opposite effect – increasing gain with increasing frequency – by swapping R1 with the combination of C1 and R2a and b. Circuits like this are the basis of *active filters,* although practical circuits are much more complex.

Op-amp integrator

If resistor R2 of Fig. 6.5 is replaced with a capacitor, the circuit becomes an integrator, Fig. 6.9. With a steady voltage on the input, the op-amp has to steadily decrease the voltage at the output to keep current flowing into the capacitor so that its inverting input stays at zero. The output voltage will be proportional to the integral of the input voltage as a function of time: the capacitor accumulates all the current that has flowed through R1 and so has a voltage which is the sum of all the voltages which appeared at the input, multiplied by the time for which they appeared. This circuit is a building block used in many complex circuits.

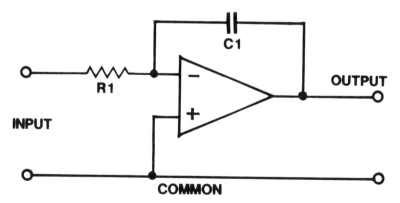

Fig. 6.9 Op-amp integrator circuit.

Voltage comparison and positive feedback

An op-amp without feedback can be used to compare two voltages, and give an output which discriminates between which is higher. This is called a *comparator*, and two projects (the Document Saver and the Pipe Saver) use op-amp comparators. Very often we have a constant *reference voltage* and we need the circuit to clearly tell us whether an input voltage is above or below this reference voltage.

In Fig. 6.10, either input 1 or 2 can be attached to the reference voltage. If input 1 is higher than input 2, the output will be at the positive supply voltage; if input 2 is higher, the output will be the negative supply voltage. However, there is a problem when the two are very nearly equal, as the output voltage will be somewhere in-between the positive and negative supplies. Because of the small imperfections and variations in the op-amp, it is impossible to tell what the voltage will be. This is not desirable; we need comparators to give a very clear indication, with the output staying in saturation at one supply voltage or the other and moving quickly between them. To get this action, most comparators use *positive feedback,* which is feedback to the non-inverting input.

For the purpose of argument, assume that input 1 of Fig. 6.11 is attached to the reference, and that input 2 is lower in voltage. The op-amp's output will saturate at the positive supply voltage, and resistors R1 and R2 will pull the non-inverting input up very slightly. For example, if the reference voltage is 0V, with the values shown this pull will be about 1/100th of the positive supply voltage.

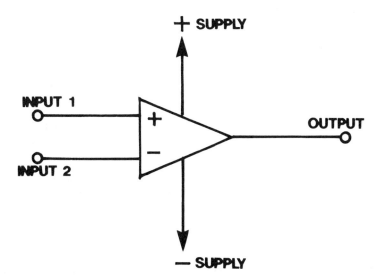

Fig. 6.10 Op-amp used as comparator.

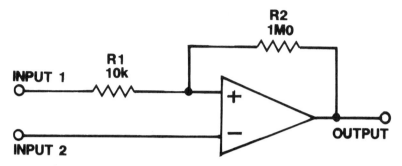

Fig. 6.11 Comparator with positive feedback added.

If the voltage at input 2 is increased, nothing will happen until it has exceeded the reference voltage by exactly the amount of the 'pull' from R1 and R2, and is equal to the voltage on the non-inverting input. The output of the op-amp will start to move away from the positive supply voltage, so the 'pull' lessens, and this makes the voltage at the output go a little lower, so reinforcing the reduction in 'pull', and so on. The net effect is that the output will switch very rapidly from saturation at the positive supply to saturation at the negative supply. If the voltage at input 2 is now reduced, nothing happens until it falls below the reference voltage by an amount equal to the 'pull', when the process reverses.

In general, as the ratio of R2 to R1 is increased, the voltage difference between the switching points in opposite directions decreases; the larger R2, or smaller R1, the smaller the difference. This circuit is said to have *hysteresis*, which means that reversing the direction of change does not take it back along the same path.

Op-amp oscillators

Op-amps can be used to make *oscillators*, circuits which produce an alternating output with no input. They use positive and negative feedback to make this happen. The simplest op-amp oscillator is shown in Fig. 6.12.

At switch-on, capacitor C1 is discharged; assume that the op-amp's output goes to the positive supply voltage (imperfections ensure that it goes to one or other supply rail). C1 charges via resistor R1, and when its voltage reaches half the supply voltage, equal to the non-inverting input, the op-amp's output voltage will begin to fall. Positive feedback to the non-inverting input via R2 and R3 makes the voltage fall further, and the output rapidly switches to saturation at the negative supply. C1 begins to charge negatively, until it reaches half the supply voltage, which is now the voltage at the non-inverting input. The positive feedback makes the op-amp's output voltage go quickly from the negative to the positive supply voltage – C1 begins to charge positively, and the whole process repeats.

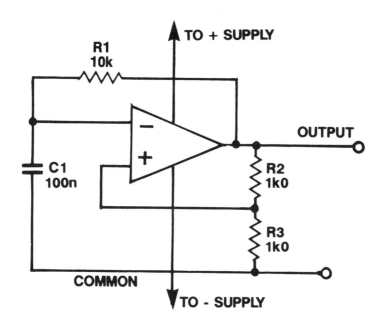

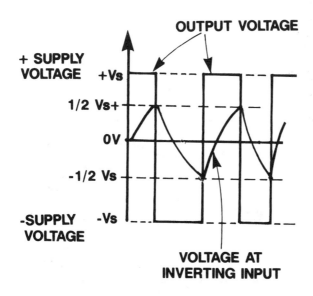

Fig. 6.12 Simple op-amp oscillator with circuit waveforms.

For the component values shown, the frequency of oscillation will be about 45Hz. R2 and R3 do not have to be the same, but when they are the period t and frequency f are given by:

$$t = 2 \times R1 \times C1 \text{ seconds}$$
$$f = 1/t \text{ Hz}$$

(R1 in ohms, C1 in farads).

When R2 and R3 are not equal, a correction factor has to be applied.

Wien bridge oscillator

This oscillator generates high quality sine-wave signals for test purposes. Fig. 6.13 shows the basic circuit; the amplifier has its gain fixed at just over 3 by negative feedback through R3 and R4. However, positive feedback also takes place through the frequency selecting network R1 and C1, limited by R2 and C2. If R1 = R2 and C1 = C2, the signal appearing at the non-inverting input of the op-amp will be in phase with the signal at the output at a frequency given by:

$$f = 1/(2 \times R1 \times C1) \text{Hz}$$

(R1 in ohms, C1 in farads).

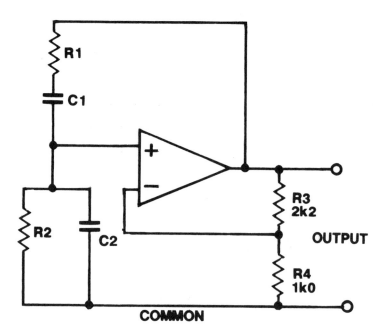

Fig. 6.13 Basic Wien bridge oscillator.

The signal at the non-inverting input will be one third the size of the signal at the output, so with a gain of just over three from the op-amp, oscillations can be sustained.

Real op-amps

One of the commonest op-amps is the 741. Many manufacturers make these, so you see the uA741, the TBB741, the LM741, etc. The 741s are always single, but the 747 is a dual version in a 14-pin package. The 741 is elderly as ICs go, but it is still used widely because it is cheap and perfectly adequate for many applications.

The pin connections for a 741 are shown in Fig. 6.14. Note that there are two more connections than already described, labelled 'offset null'.

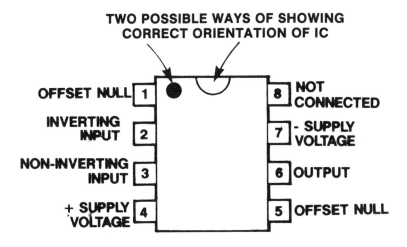

TWO POSSIBLE WAYS OF SHOWING CORRECT ORIENTATION OF IC

OFFSET NULL	1		8	NOT CONNECTED
INVERTING INPUT	2		7	- SUPPLY VOLTAGE
NON-INVERTING INPUT	3		6	OUTPUT
+ SUPPLY VOLTAGE	4		5	OFFSET NULL

Fig. 6.14 741 connections, viewed from above (i.e. pins facing away from viewpoint).

Offset and input bias

Ideally if the two inputs of an op-amp are connected together, the output should always be zero. In practice, as we have already mentioned, small imperfections in the input stages, combined with the extremely large gain, mean that the output will be in saturation at one of the supply rails, unless negative feedback is used. The *input offset voltage* is the voltage needed between the inputs to make the output zero, with no overall feedback. The input offset voltage varies from individual op-amp to op-amp of the same type, so data sheets normally cite the maximum input offset voltage.

Some op-amps have special offset null connections to allow the input offset voltage to be adjusted to zero; the circuit used for the 741 is shown in

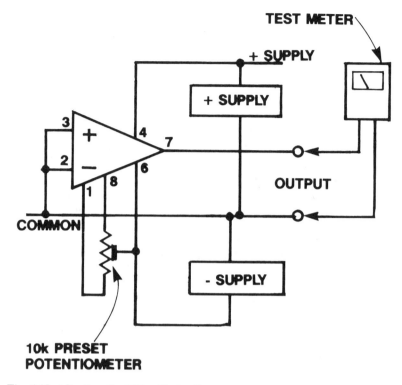

Fig. 6.15 Adjusting the 741's offset voltage.

Fig. 6.15. This adjustment is rarely necessary, only when dealing with very small DC voltages.

The *input bias current* and *input resistance* have to be taken into account when designing circuits. Op-amp inputs are connected to either the base of bipolar transistors or the gate of a JFET or MOSFET on the silicon chip. Bipolar transistors need some current flowing into them to make them work; this is the input bias current. Although FETs and MOSFETs take negligible current, there will still be a resistor between the input and the common line. To avoid problems we must make sure that currents flowing round the inputs are large enough for the input bias current and resistance (0.2μA and 1M0 on the 741) to be insignificant.

Frequency response

Two different effects limit the frequency response of op-amps. The very large open gain of op-amps begins to fall off at surprisingly low frequencies, as shown in Fig. 6.16. If you design an op-amp circuit to have a gain of 100 at 1MHz, it won't work! The *unity gain bandwidth* is the

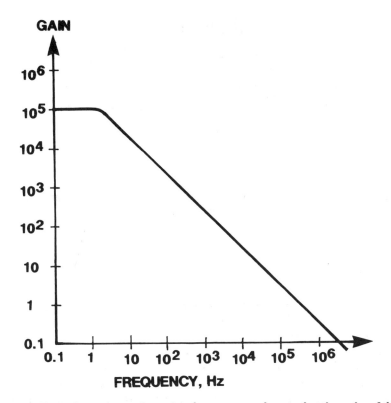

Fig. 6.16 Typical open loop gain against frequency graph; note that the scales of the gain and frequency are logarithmic.

frequency at which an op-amp's open loop gain drops to 1; although this does not give an exact figure of the op-amp's performance at lower frequencies, it is a good indicator of capabilities.

The second effect is *slew rate limiting*. The slew rate is the maximum speed that the op-amp can change its output. The 741 has a slew rate of $0.5V/\mu s$, meaning that the fastest its output can change is 0.5 volts per microsecond. Fig. 6.17 shows two signals of the same frequency; the smaller never reaches the maximum slew rate, while the larger has its waveform severely distorted.

Imagine applying any alternating input which is just large enough to produce the maximum undistorted output without clipping. If we vary the frequency, we eventually will come to a frequency where the slew rate is distorting the signal, and we must either reduce the frequency or the signal size to get an undistorted output. The range of frequencies, beginning at zero frequency, over which we can get maximum undistorted output is called the *full power bandwith.*

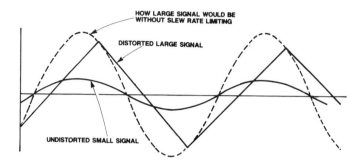

Fig. 6.17 Slew rate limiting can severely distort one signal but leave another of the same frequency untouched; it's all a matter of size.

Modern op-amps

Currently, the highest widely available unity gain bandwidth is 15MHz, and most are 1 or 3MHz. However, many modern op-amps have full power bandwidths of 100kHz or more, and slew rates are often 10V/us or better.

Modern op-amps also improve on the 741 in terms of the noise (unwanted small signal variations) which they introduce into the signal. They also improve input offset voltage and bias current, but there is a trade-off between these two. FETs (especially MOSFETs) give very high input resistances in the region of teraohms (1,000,000MΩ) and input currents of a few tens of picoamps. However, their voltages are not as predictable as those of bipolar transistors, so the input offset voltages cannot be precisely controlled. Input voltages of FET op-amps are typically worse than those of the 741, at 5mV as opposed to 1mV.

Op-amps come in packages usually with 8 or 14 pins, and in two rows 0.3 inch (7.5mm) apart, with adjacent pins 0.1 inch (2.5mm) apart. In an eight-pin package, there can be one or two op-amps; with one op-amp, the pin connections are almost invariably the same as the 741's, except that the pins used for the offset null may be used for something else or not at all. Connections to the commonest packages for dual (two) op-amps and quad (four) op-amps are shown in Fig. 6.18. Note that the power supply connections are shared.

You can use virtually any op-amp for any application, though it may not give the optimum performance. The main limitation is often supply voltage. If a circuit uses a very low supply voltage (5V or under) many types of op-amp will not work. Conversely, op-amps designed to operate on very low supply voltages may not survive a more usual supply voltage, say ±12V.

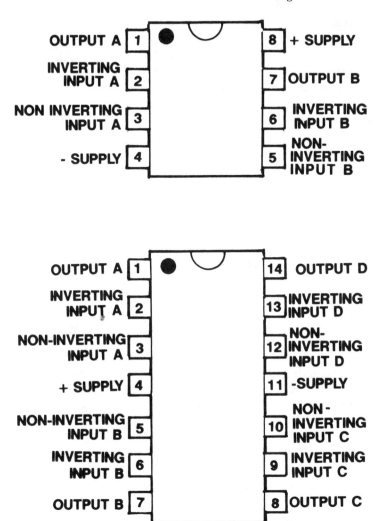

Fig. 6.18 Connections to common dual and quad op-amps—note that the connections are again viewed from above.

Audio amplifiers

Many op-amps are suitable for use in preamplifiers; there are also special purpose devices for this job. More interestingly, there are special purpose IC power amplifiers which take the 'halfway' signal from the preamplifier and produce voltages and currents large enough to drive loudspeakers. There are two main types of IC power amplifiers.

One type (like the TDA2030 used in the Stereo Amplifier project) is basically a 'super' op-amp with gain determined by feedback. The other type (for example, the widely-used LM380) has a fixed gain; although it has non-inverting and inverting inputs, the signal is applied to one and the other is attached to the common line.

Voltage regulators

A regulated power supply gives a constant voltage free from large fluctuations in output even when varying currents are drawn from it. There are two main types: *fixed voltage regulators* supply just one voltage with minimal extra components; while *variable voltage regulators* will supply a range of voltages using a few extra components, usually including a potentiometer to adjust the voltage.

The most common of the fixed type are the 78xx and 79xx series regulators. The difference is the polarity – the 78xx series are positive regulators and the 79xx are negative. The final two figures of the IC serial number give the output voltage, so the 7805 is a positive 5V regulator, the 7915 is a negative 15V regulator, and so on. There are 5, 12, and 15V versions in both series, and versions with maximum currents of 100mA, 500mA, 1A, 1.5A and 2A (although not all combinations of current and voltage are widely available). The main series are all 1A regulators, so 7812 is a 12V regulator with a maximum current of 1A. When the letter L is inserted in the middle the maximum current is only 100mA and the case style is different (eg 78L12 is +12V 100mA); M indicates 500mA maximum current (eg 79M15 is –15V 500mA max); and S indicates 2A maximum current (eg 78S05 is +5V 2A max). Details of the 78xx and 79xx series are shown in Figs 6.19 and 6.20, but note carefully that the connections to the two series are quite different.

The regulators are intended to clean up large amounts of ripple, but the input voltage to them must not dip below the minimum input level, which is 2 or 3V above the output voltage, depending on the exact regulator. If this happens, the regulator momentarily stops working, producing a surprisingly large dip in the output voltage – so do the ripple voltage calculations outlined in Chapter 4.

Regulators often need a heat sink. If a 7812 supplied from an average voltage of 18V is passing 1A, there must be 6W of power being turned into heat, which is enough to make the regulator get very hot. The metal tag can simply be bolted to any convenient piece of metal, but it must be at the same voltage as the tag; alternatively an insulation kit can be used to allow the heat to flow away without any electrical connection. These ICs regulate their outputs to within a few percent and reduce ripple to millivolts. They can sometimes oscillate, so don't leave off the 100n capacitors shown on the circuit diagrams.

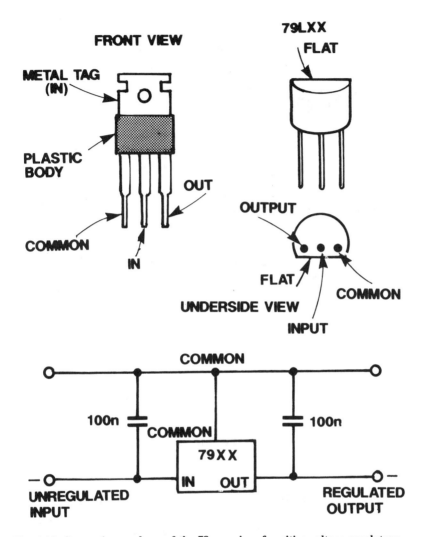

Fig. 6.19 Connections and use of the 78xx series of positive voltage regulators.

Adjustable regulators are also available, and the most common is the LM317 (Fig. 6.21), which comes in four versions: LM317L can regulate 100mA, the '317M 500mA, and the '317T and K both 1.5A. The K version regulates its output more closely than the T version, and also can dissipate more heat. By adjusting R2 in Fig. 6.21, which could be a potentiometer, the output voltage can be varied between 1.2 and 37V, provided the input voltage is at least 3V above the output voltage.

A new type of voltage regulator is currently becoming increasingly common, called the *switched mode regulator*. This uses high-frequency

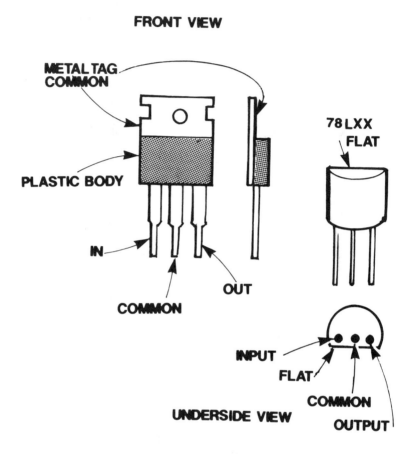

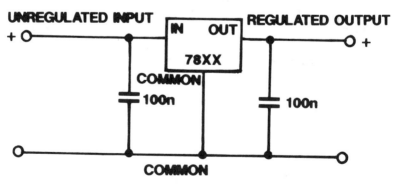

Fig. 6.20 Connections and use of the 79xx negative voltage regulators.

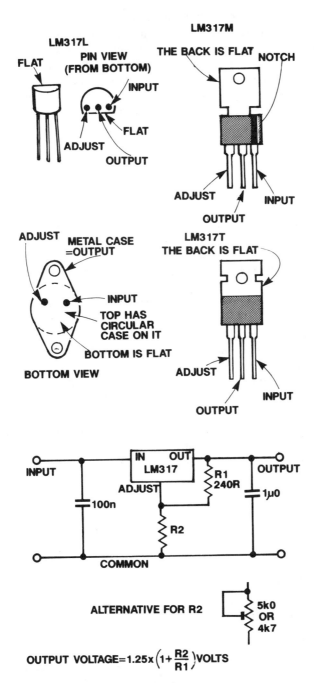

Fig. 6.21 LM317 variants and typical circuit.

switching in combination with an inductor to step-up or step-down voltages, so there is much less restriction on input voltage. They also waste very little power in the form of heat. Currently, however, they have some drawbacks. They are still much more complicated than the regulators previously described and much less standard, and they are not really suitable for audio circuits. Also they need a very precisely wound inductor. However, versions as standard as the 78xx and 79xx series should soon be widely available.

The 555

The 555 is a timer and oscillator. Fig. 6.22 shows the connections to the IC; Fig. 6.23 shows its internal circuitry, along with a few extra external components which make it act as a monostable – a circuit which has a stable 'rest' state where it can stay indefinitely, and a temporary state which lasts a certain limited period after it is set. The Alarm Controller project uses a 555 to make sure that the siren will sound for a maximum of 20 minutes, then reset (this is a legal requirement).

The 555 contains a voltage divider, two comparators, a flip-flop (described in Chapter 8), a transistor, and an output driver stage capable of sinking (taking in) or sourcing (giving out) a current of up to 200mA. The circuit works as follows: initially the capacitor is discharged and prevented from charging by the transistor which takes all the current from R1. The transistor is controlled by the output from the flip-flop which is positive at this point, the output driver buffers this and is also positive. This is the stable state of the circuit, and it can stay like this indefinitely. However, if a negative input pulse is applied to the trigger, comparator 2 switches its output, and the flip-flop stops the transistor conducting and also takes the output negative. The circuit goes into its temporary state. Capacitor C1 can charge up through resistor R1 until its voltage reaches the same voltage as

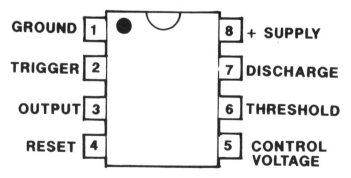

Fig. 6.22 The 555 connections, as viewed from above.

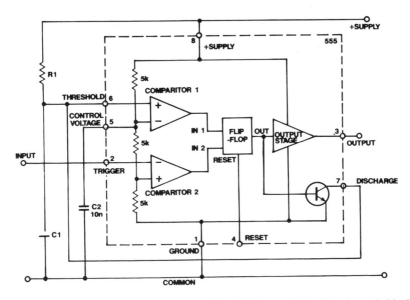

Fig. 6.23 The 555 connected as a monostable, with internal circuitry shown in block form.

the -ve input of comparator 1 (set at 2/3 the supply voltage by the three 5k resistors). Comparator 1 takes its output positive, making the flip-flop change its output. This makes the transistor conduct and discharge capacitor C1. The circuit returns to its stable state and the output returns to positive. The time (T) that the circuit can stay in the temporary state is controlled by resistor R1 and capacitor C1 and is given by $T = 1.1 \times R1 \times C1$, where T is in seconds, R1 in ohms and C1 in farads.

Fig. 6.24 shows the 555 connected as an astable oscillator (astable means it does not have a stable state, so it alternates between two unstable states). The main difference from the monostable is that the circuit now triggers itself. Resistor R2 has to be added so that the discharge takes a finite amount of time – otherwise the transistor would discharge the capacitor, and the circuit would spend next to no time in this state. However, the 555 spends longer charging up the capacitor (via R1 and R2) than it does discharging it (via R2 alone) and is said to have an uneven duty cycle. The frequency of oscillation is given by:

$$f = \frac{1.44}{(R1 + 2 \times R2) \times C1}$$

where f is in Hz, R1 and R2 in ohms and C in farads.

The monostable and astable circuits are the two most usual ways of using the 555, but it is a very flexible IC that can be used in very different ways.

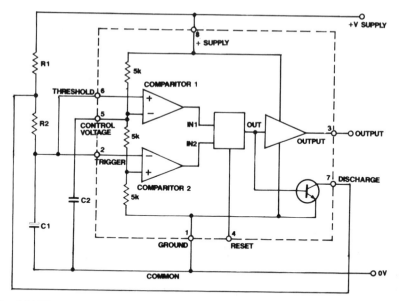

Fig. 6.24 The 555 connected as an astable.

Other linear ICs

There is a large number of types of linear IC beyond those already mentioned. However, they are generally much more specialised than the ones we have looked at. One of the larger suppliers' catalogues will give you an idea of what other types are in common use.

PROJECT: Stereo amplifier

This project uses the mains, so is not recommended as a first project.

This project is a straightforward application of two linear ICs – the TL072 op-amp from Texas Instruments and the TDA2030 audio power amplifier from SGS. It gives 8W per channel, which is sufficient to fill a medium-sized room. However, it is not recommended as a first project because it uses the mains.

I have been using it as an alternative to my normal hi-fi for two years, and it gives a very good account of itself. Besides the on/off switch, there is just one control – the volume. The 'balance' and 'tone' controls on my normal amplifier remain untouched for long periods, so I did not include them. The amplifier is small enough to fit inside the base of many record decks, avoiding one of the biggest problems of a home-made audio – making a case which is aesthetically acceptable to other members of the household.

The only potential difficulty to overcome is finding room in the case for the mains transformer.

Circuit
The circuit of just one channel of the amplifier, along with the full circuit of the power supply, is shown in Fig 6.25. (With circuits of stereo amplifiers and the like, it is usual to show just one channel if the other is the same.) The circuit is in three sections: the preamplifier, around IC1a, which takes the output from the pick-up, amplifies it and corrects for frequency response; the power amplifier, which makes the signal large enough to drive a loudspeaker; and the power supply.

In the preamplifier, one op-amp from the dual TL072 op-amp is used for each channel. This corrects for the recording pre-emphasis, where high frequencies are boosted relative to the bass frequencies following the RIAA (Recording Industry Association of America) standard. There are two sets of frequency selective components in the feedback loop – C3 and R6, and C4 and R5. The values of these components are selected to meet the RIAA requirements. Ideally, there should be no DC voltage present at the output of the preamplifier, but op-amps are not perfect so blocking capacitors C5/6/7 are used to avoid any DC being passed on. A single electrolytic capacitor used here would have an even chance of being the wrong way round if there is any DC voltage, so two are used back-to-back. C5 is added to compensate for the small inductance electrolytics have, which can affect higher frequency audio signals. It is increasingly common to use unpolarised capacitors in parallel with electrolytics in audio circuits.

The power amplifier IC2 is of the 'super op-amp' type. For audio frequencies, its gain is fixed at 22 by feedback resistors R7 and R8. For DC, capacitors C8/9/10 disconnect R7 and turn the circuit into a voltage follower; this ensures that there can only be a tiny DC voltage across the output – loudspeakers can be damaged by DC voltages of around a volt. Capacitor C13 and resistor R9 prevent the amplifier from turning into an oscillator which can happen if it is presented with a very reactive load, such as a typical loudspeaker!

The power supply uses a very simple regulator circuit, but this part of the circuit is largely responsible for giving the amplifier a respectable sound. Because the amplifier will occasionally draw currents over 2A, 78xx and 79xx regulator ICs could not be used, and higher current regulators are rather expensive. R10 and ZD11 provide a stable 15V, provided the voltage on the smoothing capacitor is a volt or two above this. C14 minimises the noise on this voltage. Q1 is a Darlington transitor connected as an emitter follower, so the output voltage at its base will be $15 - 1.2 = 13.8V$, whatever current is asked of it. An identical circuit (though of reverse polarity) takes care of the negative supply. The circuit of the transformer, bridge rectifier and smoothing capacitors C16 and 17 is a standard arrangement for a dual supply.

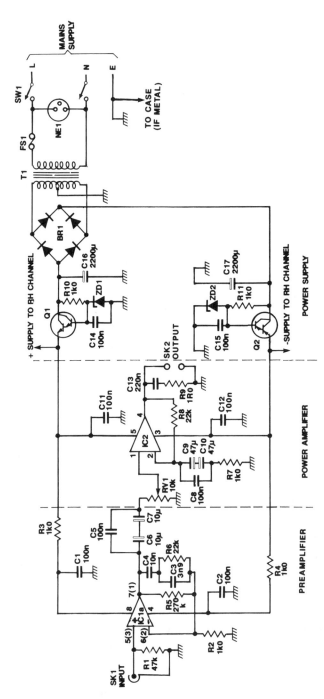

Fig. 6.25 The circuit of the audio amplifier. The circuit of the other audio channel is not shown, but the pre and power amplifiers are identical (all the component numbers have 100 added to them to distinguish them, except SK1 and IC1 which are split between the two channels).

Construction

Care must be taken in the construction of this project because it uses the mains. If basic rules are not followed, there is the danger of electric shock. Particular attention must be paid to the housing of the project and the wiring of the mains part of the circuit.

In the prototype, the amplifier and transformer were built into the base of a record deck; alternatively, the amplifier can be housed separately, and I would recommend using a metal case to help with heat dissipation and to provide an earthed container for the circuitry. Make sure the case you choose is large enough to contain all the components comfortably, with room to carry out the inter-wiring. Mounting the volume control (RV1a and b) on the PCB simplifies construction, but may not always be practical. Mounting it off the PCB allows other inputs to be connected, such as a cassette deck or tuner. The extra circuitry needed to do this is shown in Fig. 6.28. Use the PCB to gauge the position of mounting holes

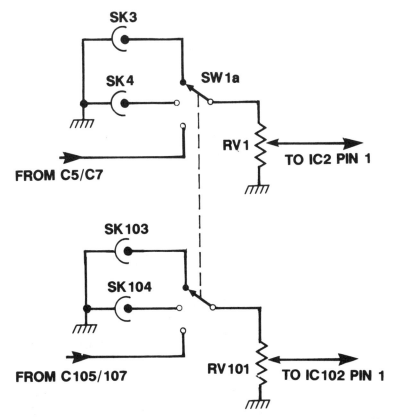

Fig. 6.26 Circuit diagram and wiring for a front panel switch which allows different inputs.

TABLE 6.1

Parts List – Stereo amplifier

Resistors (all ¼W or more, 5% or better)

R1, 101	47K
R2–4, 7, 102–104, 107	
R5, 105	270k
R6, 108, 106	22k
R9, 109	1R0
RV1/10	10k dual (stereo) potentiometer, logarithmic track

Capacitors

C1, 2, 5, 8, 14, 15, 105, 108	100n 15V or more working voltage, any sort that will fit the PCB
C3, 103	3n9 5% or better tolerance, preferably polyester or polystyrene
C3, 103	but a tolerance of 5% or better is most important
C4, 104	10n 5% or better (same comments as for C3)
C6, 7, 9, 10, 106, 107, 109, 110	10µ 25V or higher single-ended electrolytic
C13, 113	220n, any sort that will fit
C16, 17	2,200µ, 25V or higher electrolytic, axial leads

Semiconductors

IC1	Tl062 dual op-amp
IC, 2, 102	TDA2030 power amplifier IC
Q1TIP122	
Q2	TIP127
ZD1, 2	15V 1W zener diode
BR1	50V 2A bridge rectifier

Miscellaneous

T1	15-0-15V (or 17-0-17V) 2A mains transformer
FS1	500mA anti-surge fuse with mount
NE1	mains neon light (or part of SW1)
SW1	double pole mains switch, to choice
SK1	stereo disc input socket, dual phono socket or similar (not needed if amplifier mounted in base or record deck)
SK2	stereo loudspeaker output connectors (or two singles)
SK3, 4, 5	stereo input sockets, if required; dual phono or similar
SW2	Input selector switch, 4-pole 3-way (if needed)

Audio screened cable; case; PCB: PCB mounting pillars; heat sink rated at 4°C/W or lower; TO66 transistor mounting kits, four off; small quantity of silicone grease (if needed); case of choice; knobs for controls; wire, solder, heat-shrink sleeving.

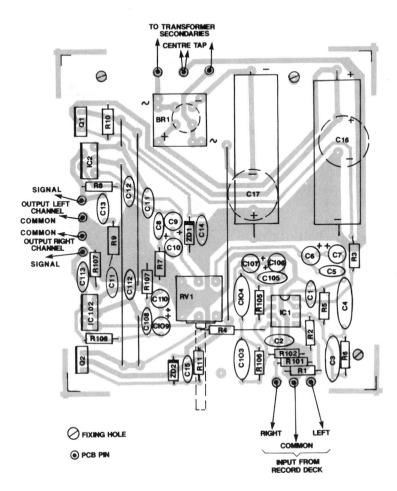

Fig. 6.27 Overlay diagram for the PCB of the Stereo Amplifier.

before assembling any components onto it. If the volume control is to be mounted on the board, the board must be oriented so that the spindle protrudes through the front panel. Plan out also the positioning of the other parts.

Assemble the PCB, inserting and soldering into position first the PCB pins, then resistors, capacitors and finally semiconductors. Do not make the leads too short on capacitors C16 and 17 because a fixing screw passes between them, or on Q1, Q2, IC2 and IC202, so you can fit them to the heat sink. Two different sizes of rectifier bridges can fit on the PCB. If the volume control is to go on the PCB, it should be the last component fitted; if it is to be mounted off-board, then pins should be soldered into its place

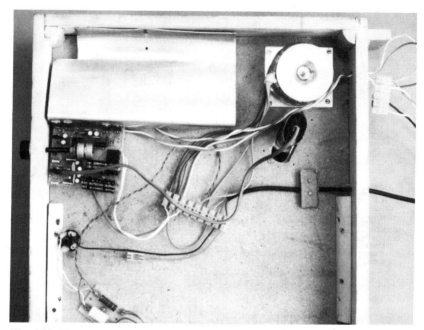

Fig. 6.28 The prototype housed in a record deck base. Note the clamping of the dark-coloured mains lead. The existing on-off switch for the deck was used for the amplifier mains supply. The sheet of aluminium (top left) was initially used as the heat sink, but one channel tended to overheat and distort; using a proper heat sink fixed the problem. The connector block near the top right is for the loudspeaker connections, and the circuitry at the bottom left is for the record deck motor and not part of this project.

for leads at the same time as the other PCB pins. Insulated leads should be used, and kept clear of other components. The long link next to C17 may need to be routed around this capacitor if an upright-mounting type is used.

Once you have completed the PCB and checked carefully for mistakes, assemble the other parts into the case. Allow good clearance for the screws used to bolt the transistors and ICs to the heat sink or metal case, as there must be no electrical contact. Special insulating kits must be used. Wire up the mains side of power supply first, keeping carefully to the circuit diagram as a mistake could be costly. In particular, check switch connections to ensure it doesn't short the live to the neutral, and double-check connections to the transformer making sure you have wired the transformer correctly. You may have to wire two separate primaries in series (for 240V operation) or parallel (if it's for 110V mains operation in the USA) – instructions should be with the transformer.

The mains lead should be firmly clamped; there are special clamps available for this, or you can improvise (on the prototype a clamp was made using two screws and some hardboard). A reasonably strong pull should not move the cable. If your transformer is a laminated type, solder wires onto the terminals to make connections. Toroidal transformers usually have flying leads, i.e. wires already connected; if these are not long enough, use a connector block designed for the mains to connect wires long enough to reach the mains switch. Cover any exposed connections to the mains live and neutral using heat-shrink sleeving. If using a metal case, attach the mains earth to the case using an earthing solder tag; this is very important.

Once the mains wiring is complete, check that none of the secondaries are shorted together or to the primaries, apply the mains and switch on. If the transformer shows any sign of distress – smoke, getting hot, continuous loud humming – switch off immediately and look for the fault. If all appears well, check the voltages on the secondaries with your multimeter switched to the appropriate AC voltage range. Switch off.

If your transformer has two separate secondaries, you need to check which way they should be connected – one way round will lead to their voltages cancelling, the other will not. Connect a jumper lead between two leads or terminals from separate secondaries, and attach the multimeter leads to the other two ends of the secondaries, then switch on. You should either have twice the normal secondary voltage or nothing at all. Switch off and disconnect the mains. If you had twice the voltage, the two ends connected together can be permanently connected to the OV connection on the PCB; if you had nothing at all, move one end of the jumper and try again. Attach the heat sink to the transistor and power amplifier ICs, using the insulator kits (leave off the silicone gel for now) and check with your multimeter switched to a resistance range that there is no electrical connection between the transistors or ICs and the heat sink. If the volume control is mounted separately attach it to the board using lengths of wire.

After checking yet again that everything is in its correct place, reconnect the mains and, keeping well clear of the board, switch on and immediately off, then disconnect the mains (there is a very small risk that the large capacitors may explode if connected the wrong way round). Using your multimeter switched to a low DC voltage range, check that there is about 0.5V across C16 and C17, the correct way round for the capacitors'

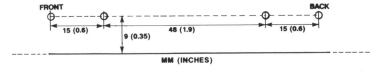

Fig. 6.29 **Position of drill holes in the heat sink, in millimetres. Holes are 5mm (or near) in diameter.**

polarities; this will drop away when the multimeter is applied. If there is no voltage, the rectifier may be wrongly connected, the fuse blown or some fault in the soldering.

If all is well so far, reconnect the mains and switch on; check the voltages across C16 and 17, which should both be about 20V (23V with a 17V transformer). Connect the multimeter negative lead to the common input at the base of the two capacitors and with the probe of the positive lead, measure the voltages around the positive regulator circuit. You should find about 15V or slightly less across the zener diode ZD1 and 13.8V or slightly under at the emitter of Q1, the output of the regulator. Attach the positive lead of the multimeter to the common point and probe around the negative regulator, checking for corresponding values.

While doing this, check on the temperatures of the power transistors and the power ICs; if they get more than warm, or if any of the voltages are seriously out, switch off immediately. Be cautious when checking temperatures, as hot metal bits can give a nasty burn. Checking around the rest of the board for overheating is also necessary. Next check the voltages at the loudspeaker outputs and at the output of the preamp, pins 7 and 1 on IC1; these should be zero or very close. If all these checks are OK, switch off, and connect loudspeakers to the outputs for them, and switch on again. There should be no sound from the loudspeakers except for very slight low frequency mains hum, which will vary according to the volume control setting. Touching one or other of the inputs to the preamp with a finger or screwdriver tip should greatly increase the hum on the channel touched. While doing this check that the power ICs and transistors do not get very hot – although it is fine if they get a little warm.

The final check before completing assembly is to apply a normal input signal to the amplifier from a record deck; the amplifier should give a good sound from the loudspeakers. Be careful not to let the power ICs or transistors overheat if you are not using the full heat sink.

Final assembly depends on the case being used. All signal wiring, except to the loudspeakers, should be done with audio quality screened cable, and any remaining wiring should be of good quality stranded lead. Some points are particularly applicable to metal cases. Make sure that none of the PCB tracks are shorted to the case. The sockets for the input and output should be insulated from the case – use appropriate versions of sockets. The circuit common has to be earthed to the case at just one point which may have to be moved to get the best results. For non-metal cases, connect the circuit earth directly to the mains earth. Make sure there are no earth loops i.e. loops of wire connected to the common line. In this project, the most likely place for this is in connections to and from the volume control and input selector switch (if fitted). The screens to these wires should be earthed by one route only. The power transistors and ICs should be mounted on the heat sink using a thin smear of silicone gel between their cases and the insulating mica washer and between the washer and the heat sink (note that some newer 'dry' washers do not need this, though). This improves the conduction of heat, but too much is worse than too little. Tighten the mounting screws firmly and, if possible, re-tighten them after several hours of use.

Fault finding

The amplifier produces a large amount of hum
Possible causes include the circuit and case being connected at an inappropriate point – usually the best point would be at the volume control – or there may be a hum loop, or a completely missed or bad earth connection, possibly on one of the capacitor leads. Alternatively, a fault which prevents one or both of the regulators working might lead to a bad hum. Solving hum problems can often involve a lot of experimentation.

The amplifier produces only limited output volume
Are you using the right sort of cartridge? The amplifier will not work with expensive moving coil cartridges, without a preamp or step-up trans-former, or with ceramic cartridges, which would probably give too great an output leading to distortion. Are the speakers unusually unresponsive – try out the amplifier with another pair if possible. Does a signal from a cassette deck, applied to the auxiliary input (or to the top of the potentiometer) give an adequate output – if it does, there is some fault in the preamplifier, if not, then the power amplifier is faulty. The gain of the power amplifier can be increased slightly by increasing R8/108 or decreasing R7/107, but this will make a relatively small difference.

The amplifier picks up radio signals
This is a very common problem. If it is just from the occasional passing taxicab or police car, it's probably not worth bothering with, but if it is from radio stations (a particular problem in summer evenings) some caution is necessary. The loudspeaker leads are acting as a radio antenna, and some part of the amplifier is acting as a radio signal detector. The standard solution is to wind both loudspeaker leads in the same direction round a ferrite ring. This doesn't always work, and again, the best solution is to experiment.

PROJECT: Document saver

This project uses the mains, so it is not recommended as a first project.
Just before a power cut occurs, the mains voltage often reduces before going off completely. A gadget to warn you when the voltage is low can be a valuable early warning if you are a computer user, and give you time to

back-up your files. The heart of this circuit is a voltage comparator, which uses an op-amp with no feedback. Other sections of the circuit are the power supply and voltage reference, a potential divider to reduce the mains voltage down to a level the circuit can deal with, and a warning bleeper which uses a piezoelectric sounder.

The power supply uses a capacitor to reduce the mains voltage; because capacitors have pure reactance and no resistance, C1 does not dissipate any heat when its impedance is used as part of a potential divider, with the combination of ZD1 and D1 the other part of the potential divider. However, the capacitor does have a very low impedance at the high frequencies which are present in voltage 'spikes' on the mains. Resistor R1 is needed to prevent these spikes blowing the fuse or overloading the rest of the circuit.

ZD1 limits the positive voltage at point A to 10V, and the diode D1 limits the negative voltage to –0.6V. D2 allows the voltage at point A to charge up C2 while A is at a positive voltage, but disconnects A from C2 while A is lower than the voltage on the capacitor.

The voltage divider for the comparator uses D3 to rectify the incoming mains voltage. R2 and R3 form the upper arm of the voltage divider, and R4 and R5 the lower. Two resistors are used for the upper arm so that if one fails, the other prevents a dangerously large current from flowing. The lower arm is divided so that the lower value resistor, R4, can be shorted out by a switch to simulate a drop in the mains voltage. The voltage given by this divider is not given by the ratio of the two resistance arms because the input is AC and the output is DC. Alternative resistor values for other mains voltages are given in the parts list; if these do not include ones for your local mains voltage, you can calculate the required value of R2 + R3 from the formula:

$$R2 + R3 = 4.05 \times V_{RMS} - 21.2$$

where V_{RMS} is the RMS mains voltage (the figure normally quoted), and the values for R2 and R3 are in kilohms. Choose resistor values from the E12 series which are similar (and above 100k) and which, when added together, come to within 25k of the exact value.

The comparator is an open-loop op-amp, with no feedback or hysteresis; one voltage comes from the potential divider described previously, the other from a simple resistive divider between the positive supply and the common line (a preset potentiometer is included so that it can be adjusted). The op-amp's output will be saturated at the common supply when the mains voltage is above a certain value.

The final stage is the bleeper which uses CMOS logic gates, described in Chapter 8. There are two oscillators, one formed by IC2a and B, and the other by IC2c and d; the first has its frequency set to about 3Hz by

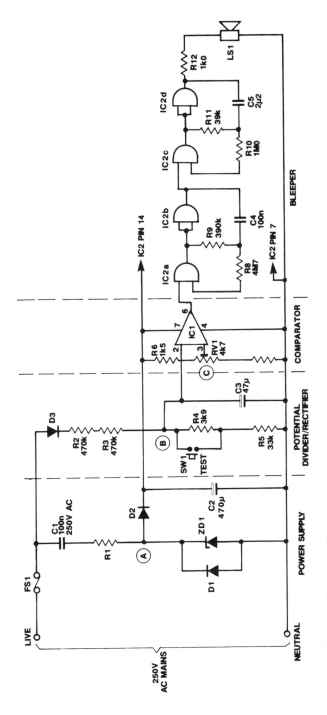

Fig. 6.30 Circuit of Document Saver.

capacitor C4 and resistors R8 and 9; the second oscillates at 3kHz, controlled by C5 and R10 and 11. The second oscillator, which drives the piezoelectric sounder, is turned on and off by the first, resulting in a bleep which is harder to ignore.

Construction and setting up
Because this circuit is live to the mains, there are a number of rules which must be kept to. They are:

* Make sure that nothing live can be touched when the box is closed and plugged in to the mains.

* Never operate the unit without a suitably rated fuse.

* Use components exactly as specified for C1, D3 and SW1.

* Where available, use a special 'plug in' case.

* If you have to test any of the circuit while it is live, keep one hand behind your back while doing so (this may sound and even look silly, but it can make the difference between a fatal electric shock and a mild one).

* It is particularly recommended that the special PCB shown in Fig. 6.30 is used for this project.

Start by positioning the fuse holder and capacitor C1 to check that they will fit in the space available; solder once positioned correctly. Fit and solder in the following order: the PCB pins for connections to the switch, the resistors, capacitors except C2 (making sure C3 is the right way round), the preset potentiometer, diodes D1–3, zener ZD1, and, taking great care to get them the right way round, IC1 and 2. Solder wires for the connections to the mains and to the switch, and connect the wires to the piezo-sounder, observing the polarity required.

Borrow a 9V battery and check that its voltage is not more than 9.5V (new batteries can have voltages above their normal rating). Connect the battery negative to the common and touch the positive terminal to point A – the unit should start bleeping. Touching the wire soldered to point B (for connection to SW1) to the common line should silence the bleeping. Switch your multimeter to the current range and connect between point A and the battery positive; the reading should be no more than 5 milliamps, with or without the bleeper going.

The next step is to fit out the case; make sure there is enough room inside to fit all the components. The prototype case was bought from Maplin Electronics and has an internal plastic frame which holds the PCB. A hole must be drilled to let the sound out, and the sounder glued over it using epoxy resin, but the hole must be so small that even a child's finger cannot touch the sounder. Drill a hole and mount test switch SW1; the connections on the back of SW1 should be covered to prevent any accidental shorts between them and the PCB. Heat-shrink sleeving was used for this on the prototype, and shrunk into position using a hair drier (a soldering iron can also be used held just underneath the sleeving, but care is needed to prevent scorching or holing it). Finally, solder insulated wires between the board and the plug pins in the special case. On the prototype case, the plug pins could be removed and held with pliers to actually make the joints.

After double-checking that everything is soldered into position correctly and there are no inadvertent shorts, assemble and screw together the unit in the case and plug it into the mains. It should begin bleeping immediately, and either remain bleeping or fall silent after 10 to 20 seconds. If it doesn't bleep at all, something is wrong, so unplug it and go to the fault-finding section.

Disconnect the unit and open the case (it will bleep for a short time after being unplugged). Select a screwdriver for adjusting RV1. If RV1 has a plastic body and plastic adjustment point, use a screwdriver with an insulated handle to adjust it. If the adjuster is metal, it is connected to the mains via the circuit so you MUST use a screwdriver with an insulated shaft as well as an insulated handle; a mains tester screwdriver would be suitable, or you could cover the screwdriver's shaft up to the tip with heat-shrink sleeving. Alternatively, cheap plastic trimming tools are available.

You must carry out the following with very great care to avoid getting a shock. Use an extension lead, disconnected from the mains. Put the top on the mains alarm's case, but don't screw it shut; plug the alarm into the extension lead, then remove the case top. Use masking tape or some other means to keep SW1's button pressed in so the contacts are closed. Check that you can adjust the potentiometer with just one hand, then put the other behind your back. Plug the extension lead plug into the mains, and the unit should bleep, possibly stopping after a short while. Adjust the potentiometer until you find the point at which the bleeper will just carry on going, i.e. it should be possible to make the bleeper carry on bleeping or go silent, if not, consult the fault-finding section. Set the potentiometer so that the bleeper just stays on, i.e. any further clockwise and the bleeping stops. Disconnect the unit from the mains and re-close the case; it is now set up and ready to use.

Plug in the unit where you will hear it, but away from devices such as washing machines and refrigerators, as these generate large 'spikes' on the mains. Pushing SW1 simulates a 10% drop in the mains voltage and will start the bleeper if the unit is functioning correctly.

Fault finding

Unit doesn't bleep when a battery is connected between A and common
Check the voltage on IC1 pin 6; if this is nearly the same as the battery

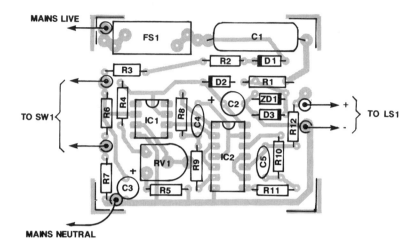

Fig. 6.31 Overlay diagram of the Document Saver PCB.

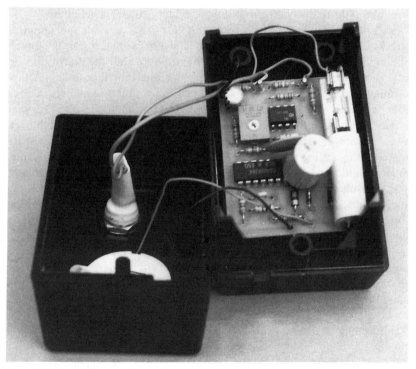

Fig. 6.32 The Document Saver in the special case.

TABLE 6.2

Parts List: Document saver

Resistors (all ¼W or higher, 5% or better; R2 and 3 must be capable of withstanding 250V)

R1	2k2
R2, 3	470k
R4	3k9
R5	33k
R6	15k
R7	10k
R8	4M7
R9	390k
R10	1M0
R11	39k
R12	1k0
RV1	4k7 sub-miniature preset potentiometer, horizontal mounting

Capacitors

C1	100n 250V AC or 750V DC
C2	470µ 16V single-ended electrolytic
C3	47µ 16V single-ended electrolytic
C4	100n 16V
C5	2n2 16V

Semiconductors

IC1	741 op-amp (uA741 or similar)
IC2	4093 (quad CMOS NAND gates with Schmitt trigger inputs)
D1, 2	1N4001 (low voltage rectifier diodes, 50PIV)
D3	1N4005 (high-voltage rectifier diode, 600 PIV)
ZD1	10V 1.3W zener diode, BZX61C10 or similar

Miscellaneous

FS1	80 or 100mA fuse, anti-surge, 200m with holder
SW1	SPST push-to-make switch, rated at 250V AC
LS1	Piezoelectric sounder (without built-in oscillator)

Printed circuit board; case; wire; heat-shrink sleeving.

Component change for 220V operation
R1 should be 390k; all other values remain the same.

Component changes for 110V operation
R1 should be 270k; R2 becomes 150k; C1 should be 220n 110V AC or 500V DC; SW1 can be rated at 110V AC; all other values remain the same.

voltage, there is a fault in EIC2 or associated circuitry. If the components are all correct, the probable cause is a damaged IC2. If IC1 pin 6 is nearly at the common line voltage, check that the voltage on pin 2 is 0V and that on · pin 3 is 3 to 5 volts. If so, and there are no other faults, ICI is faulty.

Bleeps don't stop when point B is connected to battery + terminal
Check voltage on ICI pin 6; if low with point B connected, then the fault is in or around IC2. If high, check IC1 pin 2 is the battery voltage; if it is, there is some fault with IC1.

Unit fails to bleep when first plugged in to mains
This is probably a problem with the power supply, and will have to be checked out with the unit plugged in to the mains, using an extension lead as previously described. Attach meter leads to the + and – terminals of C2 (soldering extra lengths of insulated wire if necessary) and measure the voltage between them with the mains applied (remember to keep one hand behind your back). The reading should be between 8.5 and 10V. Disconnect the mains and check the fuse, the orientation of the diodes and ZD1, and C2.

Unit carries on bleeping whatever setting of RV1 or will not start bleeping when adjusted
Attach multimeter leads between the common line and point B, and check the voltage with the mains applied (using all the usual precautions); it should be close to 4.1V, or 3.9V with SW1 closed. If not, disconnect from the mains and check the values of the resistors R2 to 5 (if the voltage is over 16V, disconnect immediately). If this voltage is OK, measure the voltage on IC1 pin 3, which should be adjustable by RV1 from 3 to 4.5V.

CHAPTER 7

Transducers: The Link to the Real World

Electronic circuits are of little use on their own. They have to be connected to the outside world to achieve anything useful – transducers and sensors make the connection.

There is an important distinction between the two. A *transducer* turns one form of energy into another. A dynamo and a photoelectric cell are input transducers because they turn one form of energy (mechanical or light energy) into electricity. A light bulb, a TV screen, and a loudspeaker are output transducers because they turn electrical energy into another form (light or sound).

A *sensor* does not convert energy but has an electrical characteristic which is affected by a physical quality (or vice-versa). A light dependent resistor, a potentiometer and a switch in a keyboard are all input sensors, because light or movement changes their electrical characteristics. Liquid crystal displays (LCDs), used in calculators and watches, are output sensors because they are either transparent or opaque depending on an applied voltage – electricity is not converted to light, and some other light source is needed for LCDs to be useful.

The remainder of the chapter takes a look at some of the commoner transducers and sensors set out according to the physical property they deal with.

Sound

Picking up and amplifying sounds was one of the first uses for electronic circuits. There are a great many different sound transducers and sensors.

Microphones can either be transducers or sensors. Nearly all use a lightweight diaphragm to pick up vibrations in the air, then use either electromagnetism, capacitance or the piezoelectric effect to convert the vibrations into an electrical signal. The most common electromagnetic microphone is the *dynamic* type, shown in Fig. 7.1. The diaphragm is attached to a coil of wire in a strong magnetic field from a specially shaped magnet and pole-piece. The coil acts like a small electric generator, so this

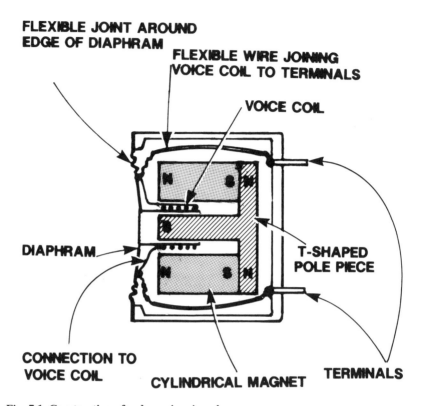

FLEXIBLE JOINT AROUND EDGE OF DIAPHRAM

FLEXIBLE WIRE JOINING VOICE COIL TO TERMINALS

VOICE COIL

DIAPHRAM

T-SHAPED POLE PIECE

CONNECTION TO VOICE COIL

CYLINDRICAL MAGNET

TERMINALS

Fig. 7.1 Construction of a dynamic microphone.

microphone is a transducer. The size of the output depends on how fast the coil and diaphragm move – we say it is *velocity dependent*. The output is typically a few millivolts.

The diaphragm of a *capacitor microphone* (sometimes called a condenser microphone) forms one plate of a capacitor (Fig. 7.2). This is attached to a high voltage source and, as the diaphragm moves in and out, the capacitance changes, causing a current to flow which can be sensed by a resistor and a built-in preamplifier. In an *electret* microphone (Fig. 7.3), the need to apply a high voltage is removed by using an electret. This is a special type of plastic which has electrical charges 'frozen' into it at manufacture, so it has a permanently fixed electrical charge across it. A preamplifier is still needed, so electret microphones usually have a small battery inside them. All capacitor microphones are velocity dependent sensors. They give outputs of several millivolts.

Piezoelectric microphones (often called *crystal* microphones) use a piece of piezoelectric material to generate their output. In basic physical construction, they are very similar to electrets, although they tend to be

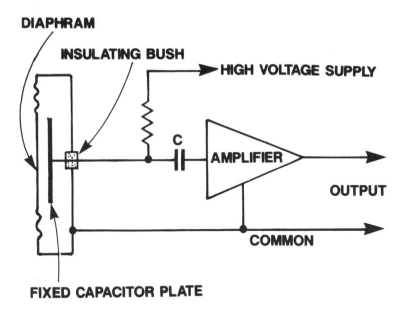

DIAPHRAM

INSULATING BUSH

HIGH VOLTAGE SUPPLY

C

AMPLIFIER

OUTPUT

COMMON

FIXED CAPACITOR PLATE

Fig. 7.2 Basic capacitor microphone.

much larger and wider. Piezoelectric substances are usually crystals, and they develop a voltage across them when compressed and stretched (they will stretch or compress in response to an applied voltage). This is a transducer, and its output depends on the amplitude (size) of the movement of the diaphragm. Output is typically around 100mV but a very high input impedance amplifier is needed.

Two other types are sometimes used. The *ribbon microphone* has a diaphragm made of corrugated aluminium ribbon held in a very strong magnetic field; sound produces a very small voltage across the ribbon, but the quality is very good and these microphones are often used in recording studios. The *carbon microphone* has its diaphragm attached to a container of carbon granules which are compressed and released by the movement, decreasing or increasing their resistance. Carbon microphones produce very low quality signals.

The direction of the sound relative to the microphone affects how well the microphone picks it up. *Unidirectional microphones* are supposed to respond only to sounds straight in front of them, while *omnidirectional microphones* respond equally to sound from all directions. In practice, unidirectional microphones respond over a range of directions and omnidirections have a 'dead' spot, from which little sound will be detected.

Loudspeakers are transducers which turn electrical variations into sound, the opposite of microphones. Construction (Fig. 7.4) is similar to a dynamic

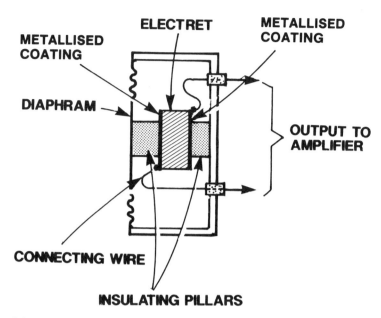

METALLISED COATING

ELECTRET

METALLISED COATING

DIAPHRAM

OUTPUT TO AMPLIFIER

CONNECTING WIRE

INSULATING PILLARS

Fig. 7.3 An electret microphone.

dynamic microphone, but with some differences to handle much larger signals. Current flowing in the movable *voice coil* causes a force between the coil and the fixed magnet, pushing or pulling the coil, depending on which way the current is flowing. The coil is attached to the loudspeaker's diaphragm (also called the *cone*) which moves forward and compresses the air in front of it, then moves back decompressing the air; at audio frequencies this produces sound waves.

While the air in front of the loudspeaker is being compressed, that behind is being decompressed, and vice-versa. The two tend to cancel each other out, which reduces the loudspeaker's sound output at lower frequencies. To avoid this, loudspeakers are usually built into a box (called the *enclosure*). Because the box is sealed, compression or decompression of the air in the box resists the diaphragm's movement. Also both the loudspeaker and enclosure have resonances (peaks and troughs in their responses) which can lead to unevenness in the loudspeaker's output at different frequencies. These factors make designing enclosures more of an art than a science.

Individual loudspeakers (often called *drive units*) cover limited frequency ranges. Hi-fi 'speakers' use two or more drive units to cover different frequency ranges – the low frequency (LF) driver or 'woofer' is for the bass, and the high frequency (HF) driver or 'tweeter' is for the treble. There may also be a middle frequency (MF) driver. HF drive units often have a dome-shaped diaphragm rather than a cone.

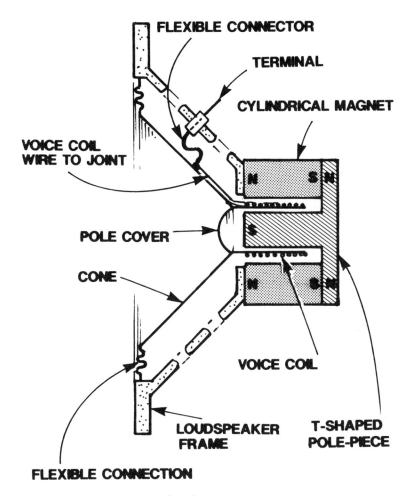

FLEXIBLE CONNECTOR

TERMINAL

CYLINDRICAL MAGNET

VOICE COIL WIRE TO JOINT

POLE COVER

CONE

VOICE COIL

LOUDSPEAKER FRAME

T-SHAPED POLE-PIECE

FLEXIBLE CONNECTION

Fig. 7.4 Basic structure of a loudspeaker.

Fig. 7.5 shows the cross-over unit used to split LF and HF signals in the Loudspeaker project. Inductor L1 in series with the LF driver allows low frequency signals through but not high; capacitor C1 restricts any high frequency signals across the LF driver still further, but R1 prevents the impedance of this part of the circuit getting too low for the amplifier at intermediate frequencies. Capacitor C2 in series with the HF driver blocks low frequency signals and lets through high frequency signals, while L2 restricts any that might get through C2. Without R2 the output from the HF driver would be too large, as it is more efficient than the LF unit. Loudspeakers have impedances which vary considerably with frequency and the impedance quoted is 'nominal' (in name only); 8 ohms is most usual, but 4 ohms is sometimes used. The maximum power handling, in

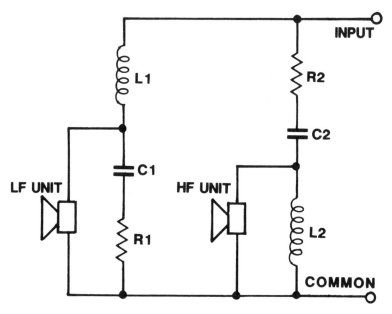

Fig. 7.5 The cross-over used in the loudspeaker project.

watts RMS with the impedance, gives the maximum voltage and currents the loudspeaker can take.

Sounders are simple loudspeakers used to give an audible signal or warning. They can either be passive, needing a signal to drive them, or active, with an internal signal source that just needs a power supply. Passive sounders, using the piezoelectric effect, are used in the Document Saver and Pipe Saver projects. Active sounders are used in the Continuity Tester and Alarm Controller projects.

Ultrasonic transducers are microphones and loudspeakers (the same unit will usually do both jobs) for frequencies well above the range of human hearing, 40kHz being typical. They have two main applications: in movement detectors and for remote control.

Heat and temperature

Finding a transducer to turn electricity into heat is no problem, as most electronic components do this all too readily. Sensing temperature is also not very difficult, because many electronic devices have electrical properties which change with temperature. However, relatively few devices offer reliable and easily usable characteristics.

Resistance probes
All resistor values vary with temperature, the extent depends on the materials used and construction. Most modern resistors offer relatively small changes, but carbon resistors give very high changes – one of the main reasons why carbon resistors are obsolete! They can be used for very low temperatures down to absolute zero, –273°C.

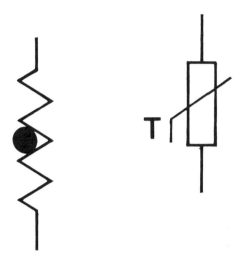

Fig. 7.6 Thermistor symbol.

Thermistors
These are resistors made from semiconductor materials, which are very sensitive to temperature. A typical thermistor might have a resistance of 5kΩ at 20°C but 200Ω at 150°C. It has a negative temperature coefficient (NTC): as the temperature goes up, resistance goes down. Positive temperature coefficient (PTC) thermistors, whose resistance increases as the temperature increases, are much less common. Thermistor resistances change very unevenly with temperature, and they are best suited to detecting whether the temperature is above or below a particular value rather than making temperature measurements. In Fig. 7.7 an NTC thermistor is used to slowly turn on the current through a transformer at switch-on. When cold, the thermistor will have a high resistance, so a much reduced current will flow. However, the thermistor heats up to over 100°C in a few seconds, and its resistance drops allowing the normal current to flow.

Zener diodes
The voltage of a zener diode varies with temperature, and this property can be used for temperature sensing. There are special purpose super-zener ICs

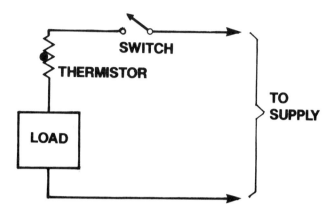

Fig. 7.7 Using a thermistor to control the switch-on current surge of transformers.

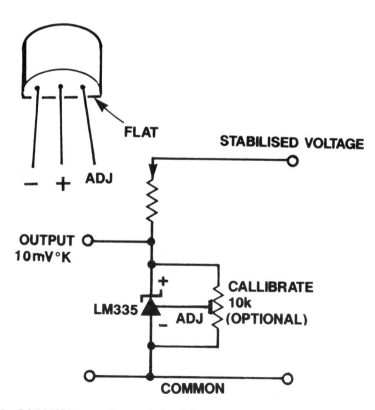

Fig. 7.8 LM355 connections and circuit for use.

where the zener voltages have been made to be directly proportional to the temperature. The LM355, shown in Fig. 7.8, is one example. The LM3911 temperature controller/measurer, used in the Pipe Saver project, is a more sophisticated device based on the same principle.

Thermocouples
If two dissimilar metals are joined together, there will be a voltage between them which depends on the temperature of the join; this is the thermoelectric effect, and devices which make use of it are called thermocouples. Fig. 7.9 shows the usual arrangement with a fixed or reference junction at a known temperature and a probe junction. The output voltage depends on the difference in temperature between the two junctions. One of the commonest thermocouples uses wires of copper and constantan, an alloy of 60% copper and 40% nickel which is used to make wire-wound resistors. This gives a voltage of about 40μV per °C temperature difference between the junctions.

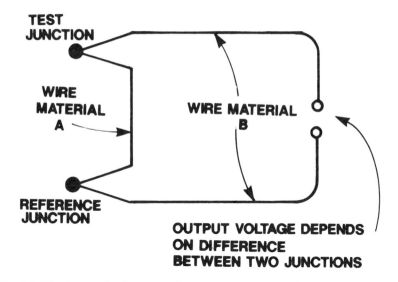

Fig. 7.9 Classic use of a thermocouple.

Light

Light is increasingly used for communicating data, for example on remote control units and fibre-optic cables which link together telephone exchanges, and recently for storing and recovering data in CD players and computer optical disk units.

One advantage of using light for communication is that there is no electrical connection, which cuts out many of the problems of circuits interfering with each other in unwanted ways. Using light in electronics has become so important that there is a special word for it – *optoelectronics.*

There are two slightly different jobs that a light transducer or sensor can do, and they require quite different characteristics. They are: accurately measuring the *total* amount of light falling; and rapidly detecting *small changes* in the amount of light. Increasingly, we need devices to do the second job, for example in CD players where small changes in the light being returned from the disk's surface are the raw signal that the player must detect and turn into the audio signal.

Photovoltaic cells
These are transducers which convert light energy into electrical energy. They may be called photocells, solar cells, or sun batteries and they have two main uses: power sources and photographic light meters. They react relatively slowly.

Light dependent resistors
These are resistors with values that decrease as the light falling on them increases. LDRs are also called photoresistors and photoconductive cells. Made from cadmium sulphide, these sensors are very slow acting and are mainly used for monitoring light levels, for example on automatic switches which turn on house lights at dusk.

Photodiodes and phototransistors
These both work in the same way. In Fig. 7.10 the junction of a diode is reverse-biased so no current normally flows. However, when a single 'packet' of light energy (a photon) is absorbed by the junction, it can release a bound-in electron and create a hole. If this occurs in the depletion zone, the electron is quickly swept into the N-type region attached to the positive battery terminal, and the hole is swept into the P-type region. Current flows through the diode, and the stronger the light (i.e. the more photons), the higher this photo-current will be. Photodiodes can react very quickly indeed.

A phototransistor takes this one stage further by making the photo-current act as the base current of a transistor, so the photo-current is multiplied by the current gain. In photo-Darlington transistors the phototransistor is the first transistor in a Darlington pair, resulting in a very high sensitivity to light. Phototransistors react very fast, but they are slower than photodiodes.

Opto-couplers
Opto-couplers allow signals to pass without the need for an electrical connection. These are useful when one part of the circuit is at mains

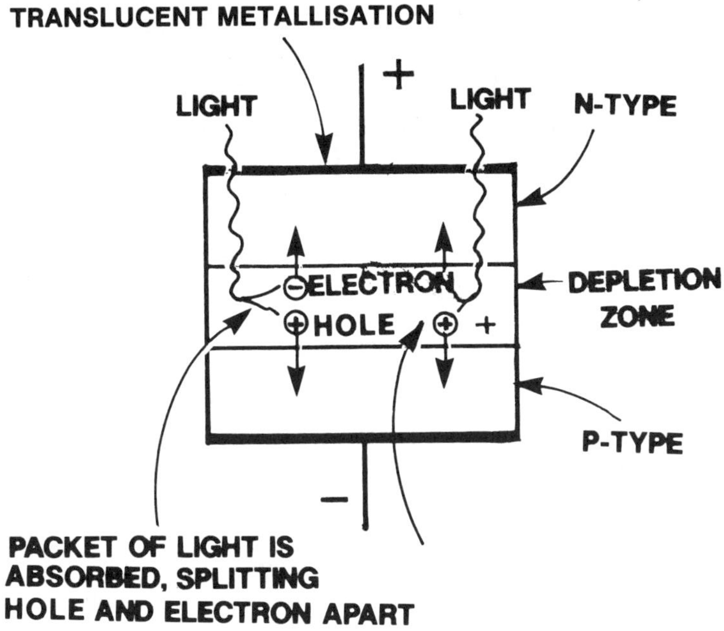

TRANSLUCENT METALLISATION

LIGHT   LIGHT   N-TYPE

⊖ELECTRON   ◄—DEPLETION
⊕HOLE ⊕ +   ZONE

P-TYPE

PACKET OF LIGHT IS
ABSORBED, SPLITTING
HOLE AND ELECTRON APART

Fig. 7.10 Photodiode structure.

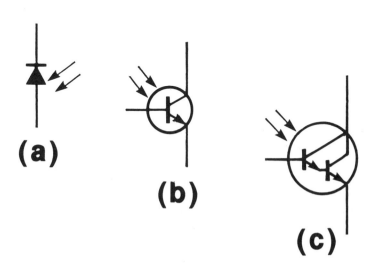

(a)

(b)

(c)

Fig. 7.11 Symbols of a photodiode, phototransistor and photo-Darlington transistor.

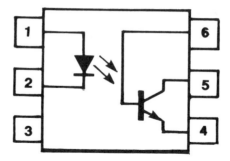

Fig. 7.12 A typical special-purpose opto-coupler.

potential, for example in a theatre lighting controller. A simple opto-coupler needs a light source (for example, an LED) and a light detector (such as an LDR) mounted in a light-tight box. For fast signals, either a photodiode or phototransmitter is needed instead of an LDR. Ready-made opto-couplers save the trouble of making your own. Fig. 7.12 shows an opto-coupler which is an LED and a phototransistor inside an IC-like package. Single and multiple opto-couplers are available, and there are opto-triacs also which can be used to control small (up to 100mA) mains currents directly – if more current is needed, the opto-triac can be used to control a more powerful conventional triac.

Optical fibres
These are thin filaments of glass or plastic which can 'pipe' light even around corners. To understand why their development has increased the importance of optoelectronics, it is necessary to delve briefly into communications theory.

To turn an audio signal into a high frequency radio signal, the audio must modulate the carrier radio signal – amplitude modulation (AM, where the audio signal controls the size or amplitude of the radio signal) and frequency modulation (FM, the audio controls the frequency, leaving the amplitude constant) are the most common methods. The result is that the radio signal no longer has one single frequency, but occupies a range of frequencies (called the bandwidth) at least as wide as the original audio signal and usually twice as wide or more. If an audio signal of 0 to 10kHz modulates a 1MHz carrier, the modulated radio signal will include frequencies from at least 990kHz to 1.010MHz, and probably wider, both in AM and FM. More audio signals can be added on the same carrier, but each will require an extra 20kHz bandwidth. The number of signals is limited by the frequency of the carrier.

The highest radio frequency currently used for broadcasting is 12GHz for satellite TV. If all the frequencies from 0 to 12GHz were used for audio radio signals, there would be room for approximately 600,000 channels, which sounds a lot until you think about the number of portable telephones, radio pagers, taxi radios etc. that exist. Many signals require much larger sections of bandwidth, for example domestic TV requires 5MHz per channel, which means there is room for 240 TV channels.

Visible light lies between 500 and 1000 THz in frequency (500,000,000 and 1,000,000,000 MHz). An optical signal has the bandwidth for 50 billion audio signals or 200 million TV signals. So a single optical fibre has the potential to carry about 100,000 times the signal as the highest frequency conventional cable. Optical fibres also lose less signal along their length than conventional cables and are lighter and smaller. At present, they are more expensive – mainly because of the optoelectronics needed for them.

Some home electronics suppliers offer a few basic optical fibre components for experimentation. A transmitter (usually an LED with a special fibre mounting), a length of optical fibre and a receiver (usually an opto-diode or transistor with a special mounting) are needed. The fibre has to couple very precisely to the transmitter and receiver, or too much light will be lost for the system to work. The light used is usually infra-red, not visible, because the optical fibre has its lowest losses in this frequency region.

Special audio and computer transducers

Two transducers are very important to audio: the magnetic tape head and the record pick-up (or cartridge); a magnetic head is also used on computer disk drives.

Magnetic tape or disks
These have areas of the surface which are magnetised to resemble lots of little magnets (Fig. 7.13). These are aligned in the direction of travel and as they pass the head, they induce fluctuating magnetisation in the magnetic head, which then induces a voltage in the head coil which recovers the signal recorded. Recording a signal is the reverse process; a signal is applied to the coil, which magnetises the tape or disk surface as it passes the gap. A very strong signal is needed to magnetise the surface, small signals will have no effect. To overcome this, a bias signal is added to the signal being recorded, so that the signal is just enough to make the difference between magnetising and not magnetising the tape or disk.

Record pick-ups
These are either crystal (piezoelectric) or electromagnetic. They work in a similar way to microphones, but rather than a diaphragm there is a lever, the *stylus*, which is tipped with a tiny diamond. The diamond follows the

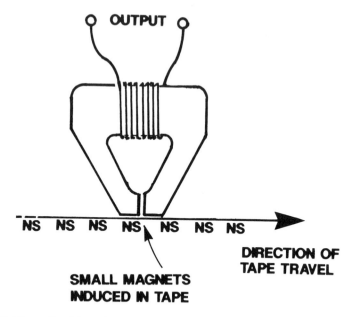

Fig. 7.13 Magnetic pick up for magnetic tape or disk.

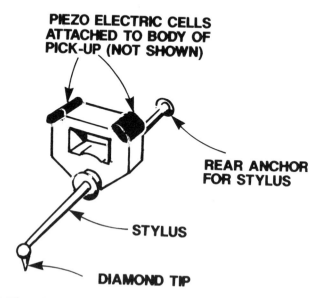

Fig. 7.14 Piezo-electric record pick-up.

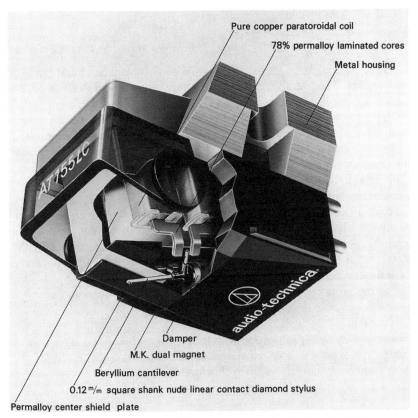

Pure copper paratoroidal coil

78% permalloy laminated cores

Metal housing

Damper

M.K. dual magnet

Beryllium cantilever

0.12 ᵐ/m square shank nude linear contact diamond stylus

Permalloy center shield plate

Fig. 7.15 Moving magnet pick-up (diagram courtesy Audio-Technica UK).

groove of the record, which has two walls, with peaks and troughs representing the audio signal.

In a crystal pick-up (Fig. 7.14), the stylus rests on two pieces of piezoelectric crystal, one for each stereo channel, which pick up the movement and convert it into electrical signals. Like piezoelectric microphones, these offer a high output (100mV to 1V) but need an amplifier with a high input impedance.

There are two sorts of magnetic pick-up. Moving magnet pick-ups (Fig. 7.15) are more common, with a small magnet attached to the stylus. Around it are coils which convert the changing magnetic field into audio signals, typically a few millivolts in size. Moving coil pick-ups reverse this situation, the coils are attached to the stylus and the magnet is fixed. The result is a lower output signal (although a few do have outputs comparable to moving coil types) but better sound quality.

PROJECT: Loudspeaker

This project is suitable for beginners. It involves only a little soldering.

One of the easiest and most worthwhile projects you can build is a hi-fi loudspeaker. The loudspeaker described here is based on a Peerless 65/2R kit, kindly supplied by Wilmslow Audio Ltd. Because getting a loudspeaker design to work well depends on a lot of skill, careful audio measurements using special equipment, and trial and error, I strongly recommend following a design from a magazine or book. Most electronics magazines and some of the hi-fi magazines publish loudspeaker designs. Also many loudspeaker suppliers have ready-made kits and designs you can follow. Some manufacturers also offer designs to follow.

To improve bass response, loudspeakers are mounted in boxes (enclosures), but this brings other problems. The loudspeaker here uses a carefully designed opening in the box to make the resonances in the enclosure and loudspeaker cancel each other out, and to reduce the resistance of the air inside the box on loudspeaker cone movement. This sort of design is called a vented, reflex or ported enclosure.

Construction
Planning is especially important with loudspeakers because it is too late to change a box once you've cut the wood and glued it together. Study the plans until you understand them thoroughly.

Fig. 7.16 Gluing together the loudspeaker panels.

If you are not using a full kit, cut the wood for the cabinet sides. Unless you are skilled at cutting wood exactly to size, I would suggest cutting a few millimetres over-size (or getting your timber merchant to cut over-size) and using a plane and shaping tool to get the dimensions exactly right, or to within 5mm at the very outside. Try assembling the boxes, swapping round sides to get the sides to meet accurately. Cut holes for the drive units, noting how they are mounted (most are now mounted from the outside). If any sides have been damaged, and these cannot be positioned where this doesn't show, fill the damaged area (this could be more convenient either before or after gluing). If you're going to paint or veneer the speakers, you could use car body filler (fibre glass) which is cheaper than special wood fillers but the wrong colour.

Assemble and glue the box, leaving one side unglued so that the contents can be fitted. If the joints are rebated on small boxes, gluing the sides to

Fig. 7.17 Holding together the loudspeakers with masking tape while they dry—not as safe as using proper woodworking cramps.

Fig. 7.18 Fitting out the inside—showing the foam linings and the tube.

Fig. 7.19 Fitting out the inside—filling up the void with absorbent material.

Fig. 7.20 Wiring up the bass and treble units.

Fig. 7.21 The final units—ready now for painting or veneering.

each other gives sufficient strength. Otherwise, glue 25mm square or triangular timber battens to line the inside of the joints, these can be tacked with small nails to keep them in position while the glue dries. Before the glue begins to dry, make sure that all joints are completely square, and that the last side will fit snugly. It is advisable to hold the joints together accurately using special woodworking clamps, but masking tape can be used to keep them in position if these are not available. Use too much glue rather than too little – the excess can be wiped away with a damp cloth while wet or sanded when dry.

Assemble the cross-overs; I suggest adding 220nF polyester, polycarbonate or polypropylene capacitors in parallel with any electrolytic capacitors in the cross-over, which marginally improves their high-frequency characteristics. When the glue is hard, fit out the box interior. Check that the drive units still fit and mark out holes for their mounting screws. LF units usually use special clamps; put these in position on the units, spacing them evenly round the edge, and mark where the holes should be drilled. Remove the drive units and drill holes just large enough to accommodate the shank of the 'T' nuts supplied with the clamps, then fit and tighten the bolts to seat them in the back of the front panel. Drill pilot holes – not all the way though – for the HF unit mounting screws.

The cross-over units can be mounted directly behind the speakers on the back panel as usual, or on the top panel, which keeps the interconnecting leads short. Solder leads onto the board long enough to connect to connectors and the drive units allowing extra length for connecting the units from the outside after the box is sealed. Check the polarities of the connections and use red wire for + and black for –. Fit the pipe for the port – its front edge should be flush with the front panel; if it is not a tight fit, glue it in position. Mount the acoustic lining on the walls, not forgetting the, as yet, unglued final side; if the lining has to be glued, wait 24 hours before the next stage, as the fumes can harm the drive units.

Bolt or screw the drive units into position, checking that the sealing rings are in position, but do not tighten fully. Solder leads from the cross-over to the drive units, checking the polarities. Be very careful not to get any solder on the speaker diaphragms, this can ruin them. Put the acoustic wadding in place, then fit but don't glue the final side; add some heavy weights (e.g. some books) to keep it in place.

Connect the speakers to an amplifier and, keeping the volume low, play some music through them. Check that all the drive units are working; it will be very obvious if there is something wrong with the bass, but you will have to put your ear close to the HF units to check they are working. If there are any problems, swap drive units and cross-overs between the speakers to find the cause. Before gluing the final side into position, remove the drive units (glue can also ruin them). Allow the glue to dry thoroughly before reassembly and use. The front of the loudspeaker can be finished off with a front covering of special acoustically transparent fabric attached to a frame of battens, or similar arrangement. The box can be painted or veneered.

TABLE 7.1

Parts List: Loudspeaker

NOTE: Exact specifications depend on design

HF and LF drive units (MF unit if in chosen design)
Cross-over networks
Acoustic lining panels
Acoustic wadding
Wood for case sides, either flooring grade chipboard or mixed density fibreboard
Mounting kits for bass units (2 off)
Porting tube (depending on design)
Wood for battens, if required (1 inch square or triangular section)
Frame for fabric cover
Acoustically transparent fabric
Connectors
Approx 2m of high-quality loudspeaker cable (or 2 × 2m each high-current single core standard wire)
PVA wood glue (Evostik, Resin W or similar)
Glue for acoustic panels, if required (Bostik No. 6 or similar).
Solder

PROJECT: Pipe Saver

This gadget is fairly easy to build, and is suitable for a first or second project. It involves soldering 16 components plus wire connections to a special PCB.

Frozen and burst pipes cause floods and loss of water supply, adding to the misery of a cold winter. This project can stop your household from suffering such a fate.

The circuit is based on an integrated circuit temperature sensor, the LM3911. There are three main sections to the circuit (Fig. 7.22): the temperature sensor; the comparator which detects when the sensor's output voltage goes below the level set by a potentiometer; and the warning bleeper. There is also a simple zener-resistor voltage regulator which supplies the sensor, op-amp and potentiometer circuits with 6.8V.

The IC's sensor is similar to a zener diode but its breakdown voltage is proportional to the absolute temperature (which has absolute zero, –273°C, as zero). At 0°C (273° absolute) the sensor's output is 2.73V; this

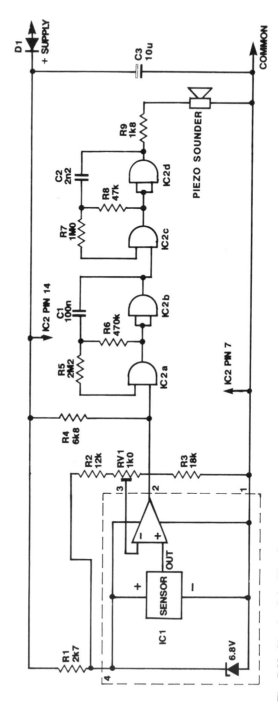

Fig. 7.22 Circuit of the Pipe Saver.

voltage appears between the + input to the sensor (i.e. IC1 pin 4) and the internal connection marked OUT which goes straight to the op-amp's non-inverting input. The op-amp can be connected either as a voltage follower (by connecting pins 2 and 3 together) or a comparator, by connecting pin 3 to an adjustable reference voltage as here. Resistors R2 and 3, and potentiometer RV1 form a voltage divider. The values are chosen so that RV1 can set the voltage between pins 4 and 3 of IC1 at 2.73V, with some room for component variation.

When the output voltage of the sensor is above 2.73V, i.e. the temperature is above 0°C, the non-inverting input of the op-amp will be more than 2.73V below the voltage at IC1 pin 4; the inverting input will be 2.73V below the voltage at pin 4, so the comparator output will be saturated at 0V, the common supply. If the voltage from the sensor is less than 2.73V, the op-amp's non-inverting input will be above the inverting input so the output will saturate; actually, the op-amp has an open collector output, meaning that its output stage has no upper transistor, so R4 'pulls' the output voltage up to nearly the supply voltage of the remaining circuitry. When this happens, the bleeper circuit is turned on. (This circuitry is the same as in the Document Saver project, so won't be described here.)

The circuit requires around 12V and 10mA, so battery powering over the winter months is not practical. Instead, an unregulated 9V 'battery eliminator' is used. The output voltage is around 12V or so when 10mA is drawn. The project's circuit is not fussy about supply voltage (though R1 may need reducing slightly if the supply voltage is below 10V), so a voltage regulator IC was not used on the prototypes. However, there is room for one on the PCB if there should be a problem. Otherwise, D1 is soldered in the position left for the regulator; it prevents accidentally connecting the power supply wrongly causing any damage.

Construction
The most temperature sensitive part of IC1 is its underside, so for best results this must be in direct contact with the pipe. On the prototype, the IC1's pins were bent back over the top, and the IC was soldered on the reverse copper side of the PCB. Note that it must be positioned the same way round as shown on the overlay. The IC can be mounted as normal on the top of the PCB, using an IC socket to lift it up, but this reduces the sensitivity.

Assembly of the PCB is straightforward; start with PCB pins, then resistors, capacitors and ICs. Link the board to the sounder, but leave off the connection to the positive connection to the battery eliminator. Plug in and switch on the eliminator and, with the multimeter on a DC volts range, check that its output is between 12 and 15V (set to the 9V range for switchable eliminators). If it is higher, put a 1k0 resistor across the output

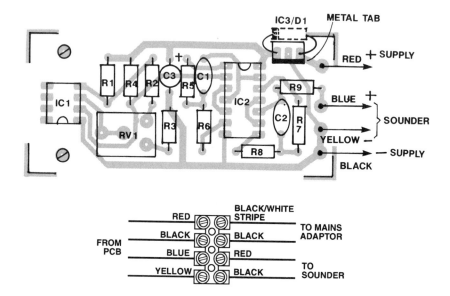

Fig. 7.23 Overlay diagram of the Pipe Saver.

and see if this reduces it to below 15V. If it doesn't, switch the adaptor to a lower voltage (where possible) or add a 12V voltage regulator to the circuit in the position provided. Connect the adaptor + lead to the PCB via the multimeter switched to a current range. The board should draw around 10mA; if the current drawn is very much more or less, switch off and look for an obvious fault.

Test the alarm by putting the entire board into the freezer compartment of a fridge/freezer (if the alarm doesn't sound after a minute, adjust RV1 fully anti-clockwise) or by using a freezing spray on IC1. Warming IC1 with a finger or with your breath will stop the alarm sounding. To calibrate the alarm, put ice and water into a metal container and stir; after 15 minutes the container will be within a degree of 0°C. Attach the alarm so that IC1 is held firmly against the container, using an elastic band for instance. Adjust RV1 to the point where the alarm sounds, but moving it slightly further would stop the alarm.

Fit the alarm board to the most vulnerable water pipe in your house, which will be the one farthest away from the warm part of the house and nearest to the roof or outside wall. You may already know from bitter experience which piece of pipe is likely to burst. On the prototype, a metal clip modified from a plumber's pipe clip keeps IC1 pressed against the pipe. Route four core burglar alarm wire from the board to the adaptor and sounder in the main body of the house. Keep the cable run to no more than 5 metres.

TABLE 7.2

Parts List: Pipe Saver

Resistors (all $\frac{1}{8}$W or more power dissipation, 5% or better

R1	2K7
R2	12k, 2% tolerance
R3	18k, 2% tolerance
R4	6k8
R5	2M2
R6	470k
R7	1M0
R8	47k
R9	1k8
RV1	1k0 sub-miniature present potentiometer, horizontal mounting

Capacitors

C1	100n, any type that will fit
C2	2n2, any type
C3	10μ electrolytic, 16V (or more) single ended (reduce to 100n non-polarised if IC3 is used)

Semiconductors

IC1	LM3911
IC2	4093 CMOS quad two-input NAND with Schmidt trigger inputs
IC3	7812 (only use if needed – see text)
D1	IN4001 (remove if IC3 used)

Miscellaneous
Piezoelectric sounder (simple type without internal oscillator); 9V DC mains adaptor unit (lower current available, typically 300mA); PCB; four core wire; solder; IC holders, if required.

Fault finding

Supply current high
Check power supply polarity (check D1 polarity too). Look for solder bridges. Check IC1, 2 are correctly oriented in opposite directions.

Bleeper doesn't sound when IC1 is frozen
Check the voltage at IC1 pin 2, which should be nearly at the supply voltage when IC1 is cooled and RV1 is fully anti-clockwise. If this is OK, there is a fault in the bleeper IC2; if not, the fault is in or around the sensor, IC1. Check first that R4 is correctly soldered.

Fig. 7.24 The Pipe Saver installed on the most vulnerable pipe in the house.

Sensor problem

Check that IC1 pin 4 is at about 6.8V; if it is less than 6.5V, decrease the value of R1 to 2k2 or 1k8, or increase the supply voltage, if possible up to 15V maximum. Check the voltage between IC1 pins 4 and 3 can be adjusted to 2.73V. Without a low reading scale on your multimeter this is difficult to see, and it may be easier to check the extreme values with RV1 at one end or the other – these should give around 2.6V and 2.85V between IC1 pins 3 and 4 (the exact values don't matter as long as they are clearly either side of 2.73 volts). If not, R2 and/or R3 will have to be adjusted. If this still hasn't fixed the problem, remove RV1 altogether and wire together pins 2 and 3 of IC1, and measure the voltage between these two and pin 4; at room temperature this should be about 3V and should decrease to 2.73V when IC1 is cooled with ice and water, or below this with a freezing spray; if it doesn't decrease, IC1 must be replaced.

Bleeper problem

This circuit is relatively straightforward, and only a bad joint, very wrong component values or a bad IC should stop this from working. If you've used IC sockets, remove IC2 and try a different IC of the same type. Static electricity can damage these ICs, so be careful to touch something earthed (e.g. a metal radiator) before handling them.

CHAPTER 8

Silicon in Action:
Digital Electronics

Digital electronics were invented to deal with numbers. Up to now our signals have been continuous, and signals could be any size, but in digital electronics signals are restricted to two possibilities. Having just two possible voltage states gives far higher reliability. The two states are:

1 *high* or *on*, where the signal voltage is at least half the supply voltage;
0 *low* or *off*, where the signal voltage is close to zero volts.

These are the signals of *positive logic;* very occasionally *negative logic* is used, where the convention is reversed and zero volts represents 1 and half supply voltage or more represents 0.

Binary

Digital electronics uses the binary system (base 2) and not the decimal (base 10) to represent numbers. In binary, zero is 0 and one is 1, but two is written 10. In decimal, 10 (try thinking of it as one-zero) means one *ten* plus zero units. In binary it means one *two* plus zero units, or 2 decimal. In decimal, 100 means one unit of ten tens (i.e. 100s), zero tens and zero units; in binary, 100 means one unit of two twos (i.e. 4s), zero units of twos and zero units, i.e. 4 decimal. Similarly, 1011 binary means one eight ($2 \times 2 \times 2$) plus zero fours plus one two plus one unit, i.e. eleven decimal.

In a binary number, all the digits are one or zero. 1030 binary has no meaning, just as the decimal number 2F3 doesn't mean anything in decimal, although it does have a meaning in base-16 (hexadecimal). Both the 3 in binary and the F in decimal are not included in the system. The highest number four binary digits can be is 1111 binary which is 15 in decimal. To go higher than 15, we just use more digits.

Bits and bytes
We use the word *bit* to mean a binary digit, a single number which can be 0 or 1. We often group four bits together to make a manageable chunk, and these are called a *byte*.

Arithmetic

Addition, subtraction, multiplication and division all follow the familiar rules, but are modified to take account of the different numbering system. Consider the binary sum:

$$0101\ 1001$$
$$+\ 0010\ 0011$$

Starting from the right, as usual, $1 + 1 = 2$ decimal, i.e. 10 binary, so 0 carry 1; next row: $0 + 1 + 1$ carried = 2 decimal, 10 binary, so 0 carry 1; next row: $0 + 0 + 1$ carried = 1 carry 0; next row: $1 + 0 + 0$ carried = 1 carry 0. The sum of the lowest four digits is 1100. Try adding the next four yourself, you should get 0111. The full result is therefore 0111 1100.

BCD and hexadecimal

Human beings find operating in binary rather painful but one way to make binary easier is to use a byte of four bits to represent a single decimal digit. The maximum value of the byte must be restricted to 1001 binary, 9 decimal. This system is called binary coded decimal (BCD). For instance, 65 decimal would be 0110 0101 BCD.

BCD is used by pocket calculators, because converting decimal to and from BCD is far easier than binary, and calculators have to convert every number keyed in or shown on their display. Doing calculations in BCD is far slower than in binary. This does not matter with calculators, because they work much faster than we can type numbers in, but it matters a great deal with computers which may perform millions of calculations in a typical program. Computers therefore do their calculations in binary, and convert the results into decimal only when it has to be passed into the human world through a display or printed output. However, humans still have to use binary numbers when designing electronic circuits and when doing some types of computer programming. The solution is to use hexadecimal ('hex'), which is base 16. We count from 0 to 9 normally, then use A for 10, B for 11, C for 12, D for 13, E for 14 and F for 15.

There is a very close correspondence between binary and hex. One byte has four digits and so a maximum value of 15, the highest value which can be represented by a single hex number. For example, the binary number 1011 0101 has eight bits and can be split into two group bytes of four bits. The first byte is 1011 binary which is 11 decimal or B hex. The second is 0101 binary, 5 decimal or 5 hex. So 1011 0101 binary is B5 hex.

If a binary number does not split exactly into bytes, add extra zeros to the left until it does (added to the left, they do not change the value), and it can then be translated into hex.

Adding electronically

To add together two bits, we need a device which has two inputs and two outputs (one for the units and one for the carry). Fig. 8.1. shows the circuit to do this, and introduces two basic logic gates. These gates are two of the basic building blocks of logic circuits; although each type does one job, the different types can be combined in an unlimited number of ways. Individual logic gates are general-purpose ICs; simple logic gates with just a few connections will have several gates on the same IC, usually all the same type; more complex gates have just one per IC.

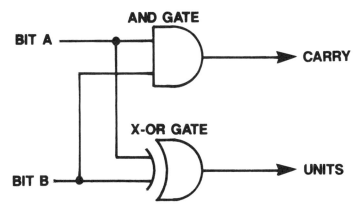

Fig. 8.1 Using an AND gate and an exclusive-OR (X-OR) gate to make a half adder.

For the simple adder shown, we require the carry output to be 1 when both the inputs are 1, and 0 otherwise. The units output must be 1 when either input is 1, but 0 if both are 1 (because a carry occurs then) or both are 0. The circuit of Fig. 8.1 is called a *half adder* because it takes no account of possible carry-in from an earlier calculation.

The AND gate in Fig. 8.1 is made so that it gives a 1 output when both inputs (A and B) are 1; if only one or neither input is one, the AND gate's output is 0. This is the function we need for the carry output.

The EXCLUSIVE-OR gate (X-OR) gives a 1 output when either input A or input B is 1, but not both at the same time; this is what is required for the unit's output.

Tables of truth

The action of any gate can be described using a table showing the possible inputs and the outputs the gate will give. This is called a *truth table*, and Fig. 8.2 shows the truth table for an OR gate. To the left are all the possible combinations of input, and on the right is the output for each combination.

INPUT A	INPUT B	OUTPUT
0	0	0
1	0	1
0	1	1
1	1	1

Fig. 8.2 Symbol and truth table for an OR gate.

INPUT A	INPUT B	OUTPUT
0	0	0
1	0	0
0	1	0
1	1	1

Fig. 8.3 Symbol and truth table for an AND gate.

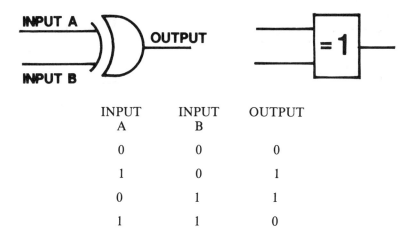

INPUT A	INPUT B	OUTPUT
0	0	0
1	0	1
0	1	1
1	1	0

Fig. 8.4 Symbol and truth table for an X-OR gate.

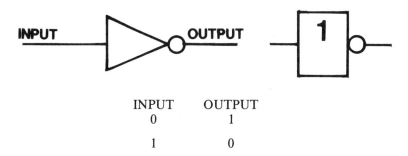

INPUT	OUTPUT
0	1
1	0

Fig. 8.5 Symbol and truth table for an inverter or NOT gate.

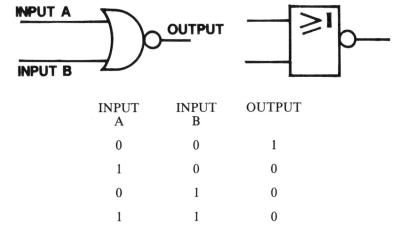

INPUT A	INPUT B	OUTPUT
0	0	1
1	0	0
0	1	0
1	1	0

Fig. 8.6 The NOR gate.

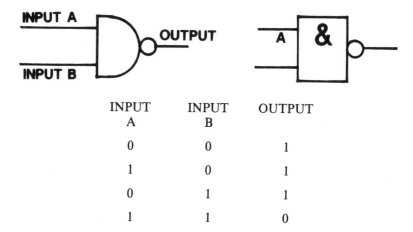

INPUT A	INPUT B	OUTPUT
0	0	1
1	0	1
0	1	1
1	1	0

Fig. 8.7 The NAND gate.

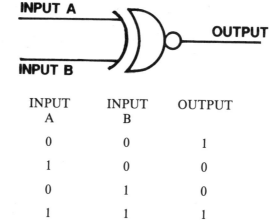

INPUT A	INPUT B	OUTPUT
0	0	1
1	0	0
0	1	0
1	1	1

Fig. 8.8 The X-NOR gate.

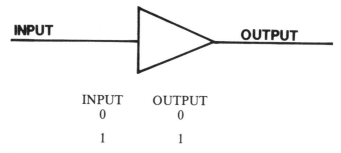

INPUT	OUTPUT
0	0
1	1

Fig. 8.9 The buffer gate.

The OR gate gives a 1 output when either A or B is 1, and gives a 0 only when both are 0. Fig. 8.3 and 8.4 give truth tables for the AND and X-OR gates that we have already met.

The inverter or NOT gate (Fig. 8.5) takes a 1 and produces a 0, and vice versa. Three more gates, the NOR, NAND and X-NOR (Figs. 8.6, 8.7 and 8.8) are made by adding an inverter to the output of OR, AND and X-OR gates. One final basic gate is the buffer (Fig. 8.9) which gives the same output as input. It is the logic equivalent of a voltage follower, and provides a high impedance input and a low impedance output.

AND, NAND, OR and NOR gates can all have more than two inputs, the only limitation being the number of pins on the IC. The rules for these gates are exactly the same, e.g. for an AND gate to give a 1 output all the inputs must be 1. However, writing truth tables for multiple-input gates becomes increasingly unwieldy; for example, the truth table of an eight-input AND gate would require 256 lines to work through all possible combinations of input. A better way is to use *Boolean algebra*.

There are three basic operators (or actions) in this algebra: the dot (.) which means AND; the plus (+) which means OR, and the bar (a straight line drawn over the top of a symbol or group of symbols) meaning NOT or inverse. Brackets are also used to group symbols and operators together, in the same way they are used in ordinary arithmetic to say 'work this bit out first'.

We can write the action of the AND gate, where the inputs are A and B as A.B; OR is A+B, NAND and NOR are $\overline{A.B}$ and $\overline{A+B}$. The eight-input AND gate mentioned earlier is described by A.B.C.D.E.F.G.H, where A to H are all inputs (not hex digits, in this case).

Rules of the game

These are some rules of Boolean algebra:

1 $1 + A = 1$ (a certainty OR a possibility, A, is always 1, a certainty);
2 $1.A = A$ (a certainty ANDed with a possibility is always the possibility);
3 $A + \overline{A} = 1$ (one or other of a possibility and its inverse must be true so the two ORed must be 1);
4 $\overline{A}.A = 0$ (a possibility and its inverse cannot be 1 at the same time, so ANDing them together always gives 0);
5 $\overline{\overline{A}} = A$ (inverting a possibility and inverting it again gets you back to where you started).

Also, in Boolean algebra, 1 is often referred to as true and 0 as false.

De Morgan's law

When designing any sort of circuitry, it is always useful to have some options open to swap it around – often the same effect can be achieved much more easily one way rather than another. De Morgan's law is a rule which allows us to design circuits to do the same job but in alternative ways. There are two versions of the law, written of course using Boolean algebra:

$$\overline{A.B} = \overline{A} + \overline{B}$$
$$\overline{A}.\overline{B} = \overline{A + B}$$

Fig. 8.10 show two sets of gates which implement the left- and right-hand sides of the first way of writing de Morgan's law, with a truth table filled showing that the two sets of gates do achieve the same result. Fig. 8.11 shows two sets of gates implementing the two halves of the second way of writing the law. Try working out the truth table yourself.

Further variations can be obtained by NOTing (inverting) both sides of the formulae:

$$A.B = \overline{\overline{A} + \overline{B}}$$
$$\overline{\overline{A}.\overline{B}} = A + B$$

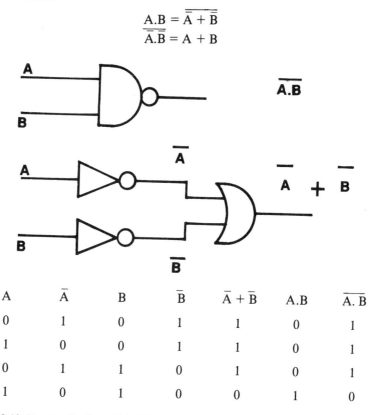

A	$\overline{A}$	B	$\overline{B}$	$\overline{A} + \overline{B}$	A.B	$\overline{A.B}$
0	1	0	1	1	0	1
1	0	0	1	1	0	1
0	1	1	0	1	0	1
1	0	1	0	0	1	0

Fig. 8.10 Proving the first of de Morgan's formulae.

and the formulae can all be extended to include three or more signals:

$$\overline{A.B.C} = \overline{A} + \overline{B} + \overline{C}$$
$$\overline{A}.\overline{B}.\overline{C} = \overline{A + B + C}$$

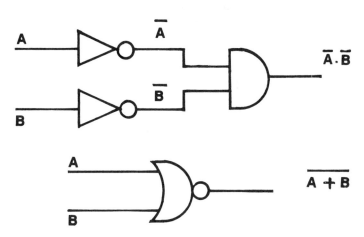

Fig. 8.11 Proving the second of de Morgan's formulae.

Using Boolean logic

The usefulness of Boolean algebra is best demonstrated with a real example. The Alarm Controller project needed to combine four signals and control a warning sounder, to indicate that you should leave promptly after arming the alarm and remind you to turn the alarm off on re-entering.

There are two alarm circuits, one delayed for exit and entry, and the other prompt protecting the rest of the house. Signals A and B from these two circuits, respectively, indicated whether they had been tripped; they are both low when tripped, but could easily be inverted to be high when tripped. Signal C is high when the alarm is set, and signal D is high during the timed exit period. I wanted the sounder to make different tones for the exit period after the alarm was first set, and for the re-entry, after the delayed circuit has tripped but before the siren is set off. For the setting tone all the outputs should be high, so the Boolean expression of this condition is:

Setting tone = A.B.C.D.

This is very easy to turn into a logic circuit using a four-input AND gate.

The warning tone should sound either when A or B goes low and C and D are both high, or when A goes low and C is high:

Warning tone = $(\overline{A} + \overline{B}).C.D. + \overline{A}.C$

This can be made easier to implement with a little algebra. Multiplying out the brackets:

$$\text{Warning tone} = \bar{A}.C.D + \bar{B}.C.D. + \bar{A}.C$$

The first and third terms can be combined, using rule 1 of the algebra:

$$\text{Warning tone} = \bar{A}.C \; (D + 1) + \bar{B}.C.D$$
$$= \bar{A}.C + \bar{B}.C.D.$$

The circuit for this (Fig. 8.12) is more complicated than I wanted. If the second term was just B.C, the circuit could be simpler, but what is the effect of making this change? D prevents the warning tone if the exit period has elapsed and the prompt circuit is tripped. This does not matter, because the main siren will be going! So we can use:

$$\text{Warning tone} = \bar{A}.C. + \bar{B}.C$$
$$= (\bar{A} + \bar{B}).C$$
$$+ (A.B).C$$

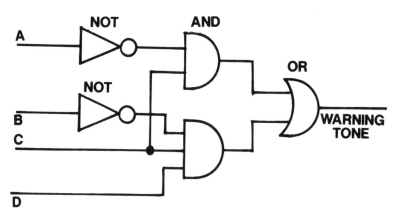

Fig. 8.12 Circuit to implement $\bar{A}.C + \bar{B}.C.D$.

TABLE 8.1

Truth table for gates of Fig. 8.12

A	B	A.B	$\overline{A.B}$	$\bar{A}$	$\bar{B}$	$\overline{A+B}$
0	0	0	1	1	1	1
1	0	0	1	0	1	1
0	1	0	1	1	0	1
1	1	1	0	0	0	0

The final version is written using de Morgan's theorem to change the OR to AND, removing the need for inverters on the A and B lines, so the circuit becomes Fig. 8.13.

The final circuit, Fig. 8.14, also includes a four-input AND gate to produce the setting tone, an oscillator giving an on-off tone to distinguish the two signals, and an OR gate which drives the sounder circuit. This has the effect that the setting tone is a continuous signal but the warning tone is an intermittent signal.

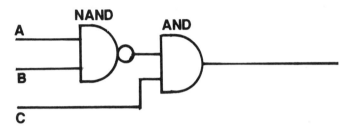

Fig. 8.13 Circuit for $\overline{(A.B)}.C$.

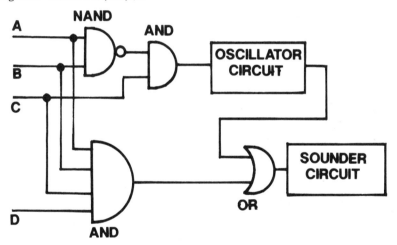

Fig. 8.14 Final circuit used.

Logic families

There are two main logic families, TTL and CMOS, and there are different sub-families within them. TTL stands for transistor-transistor logic; TTL uses bipolar transistors. CMOS stands for complementary metal oxide silicon; CMOS has N and P-channel MOSFETs on the same IC. In use, there are three main differences:

1 TTL consumes more current than CMOS, typically 2mA while quiescent (i.e. doing nothing) compared to a microamp for a similar CMOS gate. Current consumption increases with speed of activity, more markedly in CMOS because it starts off so low.

2 TTL is generally faster, though CMOS is catching up. A standard TTL gate is 10 times faster than an equivalent 4000 series CMOS gate. 74HC and 74HCT series CMOS are as fast as TTL, but currently more expensive.

3 TTL must have an accurate 5V power supply; less than 4.75 or more than 5.25 will prevent it working properly. Simple CMOS gates are much less fussy than similar TTL, for example the 4000 series CMOS will work from 5 (a little below at a pinch) to 15V, provided all the logic circuits connected together share the same voltage. However, increasingly microcomputer gates are CMOS, and these are fussy about the supply voltage.

TTL

There are several versions of TTL, which share the same pin connections and type numbers beginning with 74. Standard TTL has been all but superseded by newer types, and is rarely used; low-power Schottky (LS) TTL has the same speed but consumes a fifth of the current; advanced low-power Schottky (ALS) TTL has twice the speed of LS-TTL but consumes half the power; and Fast (F) TTL is generally the fastest TTL but has slightly lower current consumption than standard TTL. The type numbers are modified to indicate the series; 7400 is a standard TTL IC, 74LS00 is low-power Schottky, and there are also 74ALS00 and 74F00. All contain four separate NAND gates and use the same pin connections (Fig. 8.15) and much of the time can be used interchangeably.

TTL gate inputs tend to go high if not connected, so need little or no current flowing into them to give a high input. However, they need pulling down to low. TTL output stages are therefore designed to sink (take in, when making low) more current than they can source (give out to make high). Many TTL gates have just one transistor in the output with its collector connected to nothing other than the output pin (these are called open collector outputs). Often these transistors can switch loads at voltages higher than the 5V supply.

The maximum permissible voltage that is guaranteed to be recognised by a TTL input as logic 0 is 0.8V. For 1, the minimum voltage which is recognised is 2.0V. Circuits driving TTL gate inputs should avoid passing any voltage between 0.8V and 2.0V, as this will lead to unpredictable results. TTL gates themselves, used properly, will supply a maximum of 0.4V for 0 and a minimum of 2.4V for 1.

The maximum number of gates that a TTL output can drive is called the fan-out of the output; typically a standard TTL gate can drive 10 TTL inputs and 74LS gates can drive around 20 74LS inputs (a 74LS gate can

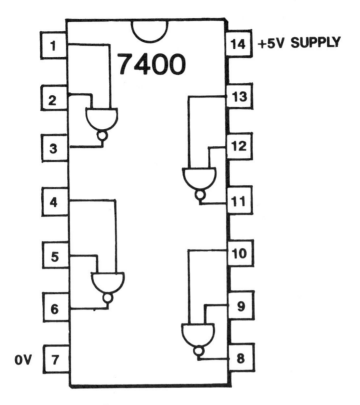

Fig. 8.15 NAND gate latch.

drive only two standard inputs). Buffers generally take less input current and can provide more output current than other gates, so increase the fan-out.

CMOS
4000 series CMOS comes in two varieties, buffered and unbuffered output. For most purposes, either type will do. A buffered output gate has an extra stage on the output which gives a better logic output but slows the gate down slightly. 4000 series CMOS has type numbers beginning in 40 or 45. Unfortunately, the type codes and pin connections are quite different from those used for TTL. For example, a CMOS 4011 has four two-input NANDs like 7400 but the connections are quite different (Fig. 8.16).

4000 series CMOS can sink and source current equally well, but provides only a milliamp or so; however, the current required by other CMOS gates is so low that this is sufficient to drive a large number of inputs. Special gates with much higher output capability are available if needed. Inter-

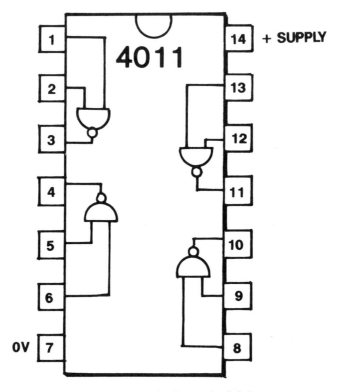

Fig. 8.16 Adding a STORE or gating facility to the S-R latch.

connecting 4000 series CMOS and TTL is tricky because the voltages for the logic levels and the input and output currents are very different, so it should be avoided.

Newer generations of CMOS use the same pin-outs and type codes as TTL with a few CMOS codes added. For example, a 74HC00 contains four NAND gates connected as the 7400, but the 74HC4002 has two four-input NOR gates like the 4002. There are two very similar series: HCT is directly connectable with TTL; and the slightly faster HC series can be connected to TTL by adding a resistor between the TTL output and the +5V supply. Both series are restricted to a 5V supply.

Latches

So far the circuits have been *decision* circuits; the output is always governed by the inputs, and changes as they change. However we also need memory circuits which can store things and act on what they have stored. This could be recalling this information or acting in a particular way depending on

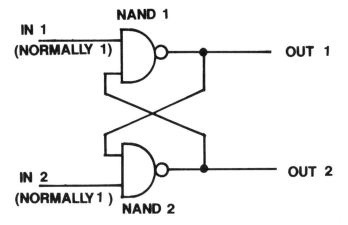

Fig. 8.17 Connections to a 7400 (or 74LS00, etc) TTL quad two-input NAND IC.

what they have stored. The names *latch, flip-flop* and *bistable* are all interchangeable terms for simple memory circuits which have two possible stable conditions (so they are bi-stables) where they will stay until moved to the other. When we need to, we can use one position to signify 1, and the other 0.

Fig. 8.17 shows a simple latch using two NAND gates. Both inputs are normally 1, but suppose IN1 is taken to 0. The output to NAND1 will be 1, because at least one of its inputs is 0, so OUT1 is 1; both inputs to NAND2 are 1, so its output is 0. If IN1 returns to 1, NAND1 still has an input at 0, so it will keep its output at 1 and NAND2 will keep its output at 0. The latch stays in this state. Taking IN2 to 0 makes its output (and OUT2) 1, so NAND1 takes its output low, so when IN2 goes to 1 again, NAND2 keeps its output at 1 and NAND1 keeps its at 0.

To summarise, while both inputs are high, the latch stays in its current state. Taking either input low makes the output corresponding go high and the other low. The latch remains in this state as long as both inputs are high. This circuit will give unpredictable results if both inputs are taken to the setting position at the same time. If inverters are added to its inputs, this circuit is a *set-reset* or *S-R latch*. The inputs are SET and RESET, and the outputs are Q and Q̄.

Clocking and the D-type latch

The simple S-R latch depends on the input lines staying at 0 unless the latch is being altered. We can use two AND gates and an extra line, the STORE or CLOCK line, to cut off the latch from the inputs until we want to change its contents. Actually, we can use NAND gates, which removes the need for

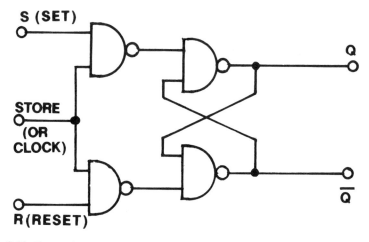

Fig. 8.18 Connections to a 4011 CMOS quad two-input NAND.

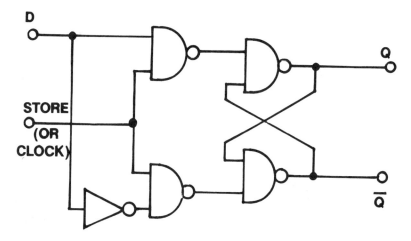

Fig. 8.19 A D-type latch.

inverters mentioned previously (Fig. 8.18). While STORE is 1, the S and R inputs can set or reset the latch as before. However, as soon as STORE goes to 0, the contents of the latch are fixed until STORE goes to 1 again. This latch is said to be gated.

A D-type latch has only one input, not two. In Fig. 8.19, the input goes to both set and reset sides, but with an inverter added to the R input. This latch holds whatever the D input was at the moment the STORE input goes to 0. To work, a D-type latch must have a STORE or clock line.

Integrated latches

The vast majority of latches are in fact integrated circuits which incorporate the gates needed to make the latch, so we do not have to worry about how the individual gates are connected, and can treat latches as just another component. Symbols for the S-R and D-type latches are shown in Fig. 8.20. IC latches are all clocked, and usually the clock is negative (latch contents cannot change while the clock input is positive); a circle on the clock input line indicates this (Fig. 8.21).

Many latches have a RESET or CLEAR input which resets the Q output to 0, whatever the state of the clock. Another possible input, PRESET, sets the latch's output to 1, regardless of the clock.

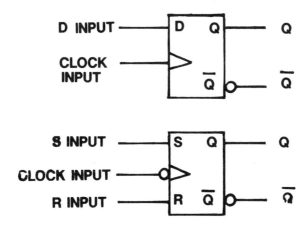

Fig. 8.20 Symbols for D-type and S-R latches.

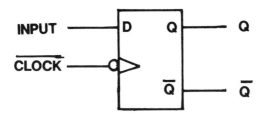

Fig. 8.21 D-type latch with negative going clock input.

Other types of flip-flop

Edge-triggered and master-slave flip-flops
With simple gated latches, you do not know what the outputs will actually be while the clock is high, because the inputs may be still changing. In complex circuits using many gates, it can take some time for the inputs to all settle down. This settling down process can be made much worse if the outputs of all the latches carry on changing too. To avoid this, these two latch types have very definite switching points at which their outputs change from one state to another (if the inputs dictate they should change).

Edge-triggered flip-flops set their contents very precisely when the gate or clock input is going from 0 to 1 (for positive or leading edge-triggered flip-flops) or from 1 to 0 (for negative or falling edge-triggered flip-flops).

Master-slave flip-flops are actually two flip-flops linked together; Fig. 8.22 shows an S-R master-slave flip-flop. The clock voltages on the flip-flops are carefully controlled and made of opposite polarities. In this example, the master latch is connected to the input while the clock voltage is 1; then the master is isolated from the input and connected to the slave while the clock voltage is 0; then the slave is isolated from the master before the master is reconnected to the input when the clock is 1 again. The output from the pair is taken from the slave.

Some flip-flops are both master-slave and edge-triggered at the same time. In most circumstances, either type can be used to achieve the same result.

J-K flip-flops
This sort of latch must be edge-triggered, master-slave, or both. J-K flip-flops are the same as S-R latches except that if both inputs are 1 a J-K flip-flop will toggle its output, so that if the output is 0 before the clock pulse, it will be 1 after (or vice-versa). The J-K flip-flops inputs are labelled J and K; J corresponds to S and K to R of an S-R latch.

T flip-flop
The T-type flip-flop has no inputs except the clock, and possibly PRESET and CLEAR. It toggles its output on every clock pulse.

Other types of gates and latches

The following is a selection of some of the commoner gates and latches not already described.

Schmidt triggers
These are devices used to 'clean up' digital signals, restoring sharp edges. A TTL Schmidt trigger will recognise an input as 1 if it exceeds 1.6V;

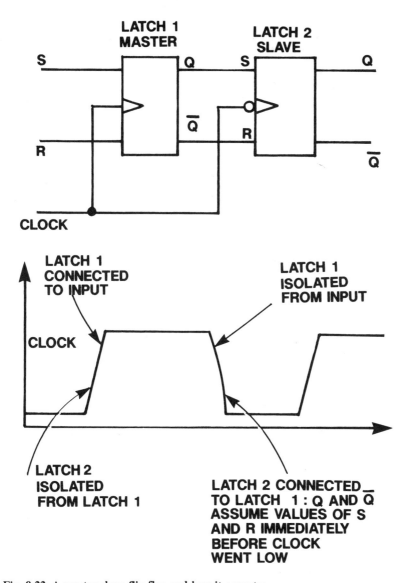

Fig. 8.22 A master-slave flip-flop and how it operates.

however, the voltage then has to decrease to 0.8V to be recognised as 0. It is said to have hysteresis, which removes any uncertainty about the point where the output will change state. Certain NAND gates (two or four input) may have Schmidt trigger inputs and some tri-state buffers are non-inverting Schmidt triggers.

Tri-state outputs

These are designed so that several different devices can output data onto the same output line. The output has a special 'third' state, where it is disconnected from the output line so allowing another output to control the line without contention (two outputs fighting each other). Most computer ICs have tri-state outputs, and there are special tri-state buffers which have their outputs disconnected unless their output enable connection is 1.

Data latches

Data latches are ICs with four or eight D-type latches all with the same control lines. They are useful for storing one or two complete bytes. These ICs often have tri-state outputs.

Counters

Counters are ICs that count in binary. Fig. 8.23 shows a simple counter using negative edge-triggered T-type flip-flops with a negative CLEAR input. Before counting, CLEAR is temporarily taken low setting all the outputs to 0. On the negative edge of the first pulse applied to the input, the output of latch 0, Q0 toggles from 0 to 1; when the next pulse comes along, Q0 changes from 1 to 0, which makes latch 1 toggle its output, Q1, from 0 to 1. The next pulse makes Q0 go from 0 to 1, which does not affect latch 1. The next pulse makes Q0 go from 1 to 0, so latch 0 toggles its output from 1 to 0 which now makes latch 2 toggle its output from 0 to 1. At this point, four pulses have arrived and the output from the latch is 0100, i.e. 4 in binary.

The count will continue until all the outputs are 1; when the next pulse arrives, all the outputs will go to 0 and the process begins again. The last output could be used to drive the input of another four-bit counter (the two counters are cascaded) to make an eight-bit counter. This is called a *ripple counter* because the count ripples through from one flip-flop to the next. *Synchronous counters* also exist, where all the outputs change together.

BCD counters can also be made by arranging the next count after 1001 in a binary counter to be 0000. There will have to be a special 'carry' output too, though there is one already on most pure binary counters.

Shift registers

These circulate the bits in a binary number. Fig. 8.24 shows a series of edge-triggered or master-slave D-type flip-flops connected to do this; on a clock pulse, each flip-flop will store the contents of the previous latch in the chain. The binary number may be fed into the shift register from the end, or there may be inputs to each latch so that they can all be loaded simultaneously. Some shift registers can be made to shift their contents in either direction.

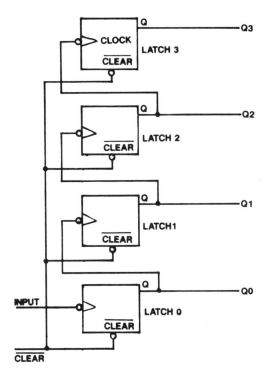

Fig. 8.23 A simple ripple-through four-stage counter.

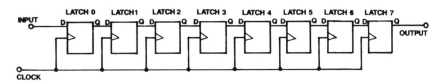

Fig. 8.24 A shift register; if desired, the output can be fed back round to the input.

Data selectors or *data multiplexers*
These have a number of data inputs, up to four address inputs and a single data output. The address input selects one of the data inputs to pass on to the output. If the address input is 1001 binary (9 decimal), the data on input 9 will appear at the output.

Magnitude comparators
Magnitude comparators typically have two groups of four inputs which are treated as binary numbers. The outputs indicate which of the two numbers is larger, or if they are equal.

Propagation delay and glitches

The measure of speed of a logic gate is its propagation delay, which is the time (usually in nanoseconds, ns) the output takes to change in sympathy with a change at the input. For example, the propagation delays of the 7400 and 74LS00 are typically 11ns and 9ns. Forgetting about propagation delays can lead to trouble when designing circuits, even ones where speed is not critical. The circuit of Fig. 8.25 should ideally always give an output of 1 because the inputs to gate IC1d must be different. However, if the output changes from 0 to 1, the top input goes from 0 to 1 immediately but it takes three times the propagation delay for the lower input to change from 1 to 0; during this time, IC1d has both inputs high, so its output will briefly go low.

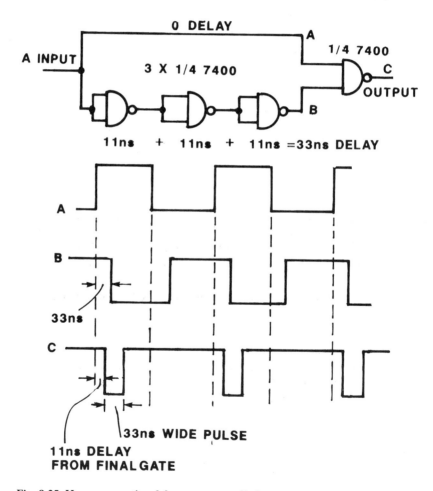

Fig. 8.25 How propagation delay can cause glitches.

This sort of problem is called a glitch, which is an unintended error pulse. Glitches cause lots of trouble and led to the invention of synchronous or clocked logic.

Synchronous logic

A synchronous system uses a master clock (a very accurate digital oscillator) to make complex logic circuits work step-by-step. Most large systems contain many flip-flops (either edge-triggered or master-slave types) which all take their inputs in at one point on the clock pulse, then set their outputs. The gates between the flip-flops can take their time to settle before the next clock pulse. Any glitches which occur between the clock pulses have no effect, as long as all the gates are settled by the next clock pulse.

Memory

Computer memories are typically ICs with many thousands of locations at which data or instructions can be stored in binary. All have data connections for the data itself, an address bus which tells the IC where to store or recall the data, and several control inputs, including the READ/WRITE line which tells the IC whether it should be storing or recalling.

A computer needs two sorts of memory. One sort to permanently contain certain instructions on what to do when first switched on and how to carry out certain regular tasks, which must not be lost when the power is turned off. The other sort is needed to temporarily store (write to memory) and recover (read from memory) the data the computer is dealing with and the programs giving specific instructions on what to do with the data; the contents of this sort must be alterable. Two types of memory IC are available: read only memory (ROM) has contents which are fixed and retained even when the power is switched off; and random access memory (RAM) can be read and written to but loses its contents when the power is removed.

There are three different types of ROM, described here in order of increasing price. Mask-programmed ROM has the data written into it at the time of manufacture. Many thousands of ROMs are made with exactly the same contents. Fusible link ROMs or PROMs (programmable ROMs) are programmed after manufacture by a special machine which applies a high voltage to memory locations, 'blowing' a micro-miniature fuse to write permanent 0s or 1s. If a mistake is made, it cannot be corrected. Fusible link ROMs have been largely replaced by EPROMs.

EPROMs are erasable PROMs. The EPROM has data written into it by applying a high voltage, a little like the fusible link ROM, but the whole of the data in the EPROM can be erased by strong ultra-violet light; it is not possible to erase just part of the data. EEPROMs are electrically erasable PROMs similar to EPROMs, but erasing is achieved by applying a high voltage to a special pin. EEPROMs are also called EAROMs, for electrically alterable read-only memory.

There are two sorts of RAM. Static RAM uses individual flip-flops, made using two MOSFETs, for each bit. Dynamic RAM uses very small capacitors to store data, one for each bit. The capacitors are less than a picofarad in size, so leakage results in the data being lost quickly unless every bit is refreshed (read and re-written) every millisecond or so. This is done by the computer itself or by special-purpose memory controller ICs. Because dynamic RAM is so simple, a large amount of memory– several million locations– can be fitted on a single IC. However, static memories are always increasing in capacity too.

PROJECT: Alarm controller

This is a difficult project, and involves a lot of careful soldering and other work.

This project is a control unit that can be used as the basis for a house alarm or car alarm. It is fitted onto a small PCB, but it has all the functions of much larger control units.

The block diagram, Fig. 8.26, shows how the different sections of the control unit work with each other and the two key-operated switches which control it: the ARM switch, which is normally used to arm or disarm the controller, and the DISABLE switch which is used when carrying out work on the unit.

To detect an intruder, the controller can use a variety of widely available sensors, from simple door-switches to sophisticated infra-red movement detectors. Sensors are attached to two circuits: the prompt circuit as used for most sensors sets off the siren as soon as it is triggered; the delayed circuit is used for sensors in the entry route, and allows you time to disarm the controller when you return. There is a simpler third circuit, labelled '24 hours', which protects the alarm itself from tampering.

Both main circuits can use sensors with normally open or normally closed contacts. Normally open switches are wired in parallel to the normally open loop, which trips if any of the switches closes. Normally closed switches are wired in series to the normally closed loop which trips if any opens. If a normally closed loop is not required, its terminals must be linked together by a short length of wire. The normally open circuits can be left unconnected. The 24 hour circuit has only a normally closed loop.

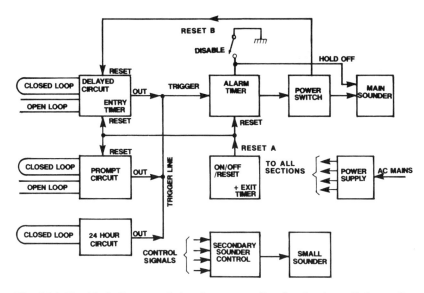

Fig. 8.26 The block diagram of the alarm controller showing how all the sections relate to each other.

The alarm timer section allows the siren to sound for approximately 20 minutes after the controller is triggered (in the UK, all alarm controllers are required to limit this time). The alarm timer drives an electronic switch, which is a medium power transistor. This can supply up to 250mA to the siren; if the siren needs more, a relay should be added. The ARM key-switch arms and disarms the two main circuits and resets the timer, silencing the siren if necessary. However, the siren can still be set off by the 24-hour circuit unless the DISABLE key-switch is used to disarm the controller completely.

The controller also has an exit timer which allows you to leave without setting off the siren. As soon as the ARM switch is set, the exit timer begins to time and the bleeper, controlled through the logic discussed earlier in this chapter, sounds and warns you to leave promptly. While the exit timer is active, only the 24-hour circuit can set off the siren.

The final section is the power supply. To save trouble, a 1A mains adaptor is used to supply an unregulated DC voltage; a 12V regulator IC is included on the PCB.

The circuit sections in detail

The delayed circuit (Fig. 8.27) uses a set-reset latch formed by a pair of two-input NAND gates, IC1a and b. When the top input to IC1a is taken low

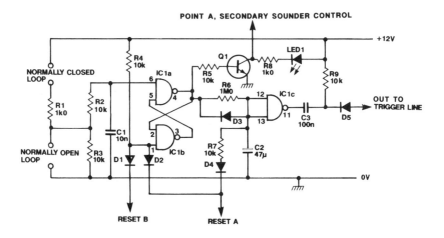

Fig. 8.27 The delayed circuit.

the latch is set. The top input goes low when either the normally closed loop is opened and R3 pulls the input low, or when the normally open loop is closed. R2 and C1 prevent any radio signals getting through and upsetting the circuitry (radio signals can be picked up on long cable runs to sensors).

When the latch is set, the output of IC1a is high. This begins to charge C2 via R6; with the values shown, it will take about 20 to 30 seconds for C2 to reach a high enough voltage to make the input to IC1c recognise it as high (IC1a to d are all Schmidt trigger NANDs), at which point IC1c goes low very rapidly sending a negative pulse down the trigger line via D5. LED1 is illuminated by current from transistor Q1, which receives base current from IC1a's output via R5. LED1 shows whether or not this circuit has been tripped.

There are connections to two reset lines. RESET A is the general reset, taking this low resets the latch as well as discharging the capacitor via R7 and D4. RESET B resets the latch while the sounder is going, but not the capacitor. This prevents the controller re-triggering if the condition which originally caused it to trigger has not been removed.

The prompt circuit (Fig. 8.28) uses two NAND gates (IC2a and b). RESET A is connected to one input to prevent the output of this gate going low and sending a pulse down the trigger line while the controller is not armed. However, if either of the two circuits are unset when RESET A is released, the siren will be set off immediately.

The 24 hour loop (Fig. 8.29) uses a transistor instead of a gate to avoid the need for another IC (the gate which would have been used here comes in fours or sixes, and only one was needed). Keeping the loop closed steals the

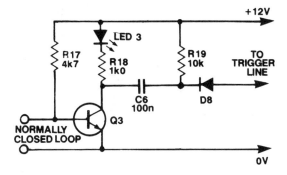

Fig. 8.28 The prompt circuit.

base current from the transistor Q3; if the loop is opened, Q3 will come on rapidly, taking its collector low and sending a negative pulse via C6 and D8 onto the trigger line.

The alarm timer (Fig. 8.30) uses a 555 timer, IC3, as a monostable. R24 and C9 are the timing components, and give an interval of about 20 minutes. Triggering needs a negative pulse from any of the circuits to take pin 2 briefly low.

The output goes to a power transistor, Q4, via a 220R resistor, which limits the current into the transistor's base. Q4 provides enough current for most sirens, but if you use a type which needs more than 250mA, it would be better to add a 12V relay between Q4 and the siren unit (put the relay's coil where the siren would normally go). The remainder of the timer circuit is concerned with resetting IC3. To get the 555 to reset, a pulse is needed to take pin 4 to zero. RESET A reaches IC3 via C7 and D9, and the latter

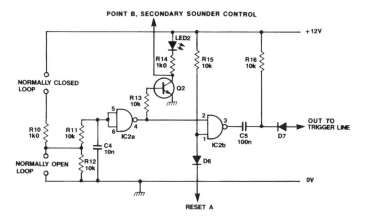

Fig. 8.29 The 24-hour loop circuit.

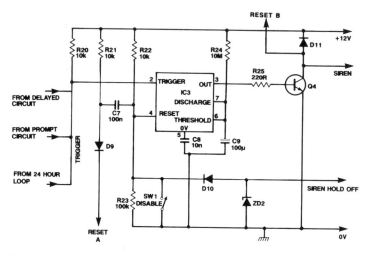

Fig. 8.30 The alarm timer.

'loses' 0.6V, so RESET A alone cannot reset IC3. To compensate for the loss, pin 4 is held at 1.1V below the positive supply voltage by potential divider R22 and R23. When a negative pulse arrives from RESET A, it only has to take pin 5 down by 11V rather than the usual 12V.

SW1, the DISABLE switch, can hold the reset permanently low, preventing the siren from going off at all – a useful facility for installation and testing. If the alarm protector is used in conjunction with this project, SW1 can also disable the case switch on the siren; D10 is needed to do this. ZD2 in this circuit is the ZD2 in the alarm defender project, mounted in the controller box.

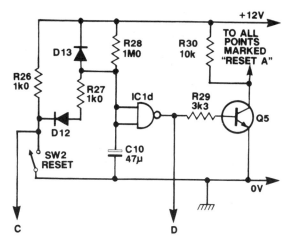

Fig. 8.31 The reset and exit timer section.

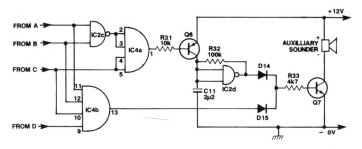

Fig. 8.32 The control logic and auxiliary sounder.

The exit timer and reset circuit is shown in Fig. 8.31. While SW2 is closed, C10 is held almost completely discharged by D12 and R27; IC1d's inputs are low, so its output is high, and current is supplied to the base of Q5 which keeps all the RESET A points low. When SW2 is opened, C10 charges through R28; it takes about 20 seconds for this to reach a high enough voltage to make the output of IC1d go low, releasing Q5 and all the RESET A points. If SW2 is closed again, C10 is very rapidly discharged through D12 and R27, taking reset A low again.

The auxiliary sounder circuit (Fig. 8.32) is the same as Fig. 8.14, discussed earlier, with a few practical details added. Since four-input NAND gates come in pairs, the second was used instead of a two-input NAND, with two pairs of inputs connected together (two inputs could have been connected to the +12V line instead if this had been more convenient). IC2d is used as an oscillator, and C11 and R32 are the timing components (this oscillator is similar to the op-amp oscillator of Fig. 6.12). If IC4a's output is low, Q6 is supplied with base current via R31 and holds C11 fully charged, stopping the oscillator with its output low. The OR gate is formed using D14 and 15; Q7 provides the current to the sounder when either input to the diode OR gate is high.

The power supply (Fig. 8.33) is a straightforward application of a 12V regulator, IC5. A mains adaptor is used to provide a source of unregulated voltage. Diode D16 is added between the regulator output and the 12V line

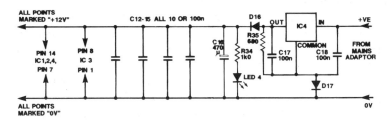

Fig. 8.33 The power supply, showing also power connections to the ICs in the remainder of the circuit.

to prevent use with a back-up battery damaging the regulator. D17 compensates for the 0.6V lost, by increasing the regulator output to 12.6V. When used with the alarm protector, the back-up battery in that project can supply the alarm controller if D2 in the alarm protector is replaced with a wire link.

Construction

A project of this complexity has to be constructed and tested in stages, and you should move on to the next stage only after the one in hand is working correctly. Always remember to remove the power by unplugging the mains adaptor while doing any assembly work, or components will be damaged.

The power supply components shown in Fig. 8.33 should be installed first; use stranded wire to connect LED4 to the board because it is mounted on the front panel of the case. All the other components of this section are on the PCB. Install all the wire links at the same time. Attach the wires from the mains adaptor, cutting off the 12V adapter plug if there is one (not the mains plug!), and check that the supply gives 12V or fairly close. IC5 needs a heat-sink; a piece of aluminium was used in the prototype; and this will be approximately 0.6V above the common voltage, unless an insulation kit is used.

Insert and solder the components for the delayed circuit of Fig. 8.27; put a short piece of wire between the NC terminals on the connector block. LED1 is connected using lengths of stranded wire. Also install R28 so that IC1d's input does not build up a static charge. Add 100k resistors in parallel with R6 and R28 on the foil side of the PCB (this reduces the timing periods to make testing much easier) and solder two short wires where the collector and emitter of Q5 will go.

Apply the power and short together the NO contacts using a short length of wire. LED1 should light immediately, and after a few seconds the voltage at IC1c pin 11 should go from approx 12V to 0V. Touching together the wires soldered at Q5 collector and emitter will reset the circuit. Check that it will trigger if the wire link in the NC circuit is briefly removed.

The timer circuit of Fig. 8.30 should be built and tested next. Add a 100k resistor parallel with R24 on the foil side of the PCB, to reduce the timer period. Connect a small 12V light bulb (3W maximum) or buzzer instead of the siren on connector block CB5 (make sure the buzzer is connected the right way round if it has + and – terminals). Apply power. Touching together the two wires at Q5's position should reset the entire circuit, including the siren (or rather, its substitute) if this starts going. Trip the delayed circuit as before by briefly shorting the NO circuit; LED1 should light, then a few seconds later the siren should go off and then stop it after about 10 seconds. LED1 should be extinguished (provided the two wires at Q4 are apart) as soon as the siren sounds, indicating that the circuit is reset.

Tripping either the NC or NO circuit should make the entire sequence repeat. Closing switch SW1 should prevent the alarm from tripping, and should cut off the alarm if it has tripped.

The prompt and 24-hour circuits (Figs. 8.28 and 8.29) can be built and tested next; also install C11 to prevent static damage to IC1d. Both circuits should trip the siren with no delay, and should be capable of re-tripping the alarm after NC and NO contacts are restored to their correct states.

Remove the wire links at Q5 and build the reset circuit of Fig. 8.30 (put another 100k resistor in parallel with R28). The RESET A line should go low as soon as SW2 is closed. After SW2 is opened, RESET A should stay low for about 3 seconds. Check that closing SW2 stops the delayed and prompt circuits from tripping the alarm, but not the 24 hour circuit. Also, shutting SW2 while the siren is going should stop it.

Finally, build the auxiliary sounder circuit of Fig. 8.32. The sounder should operate continuously when SW2 is opened until RESET A goes high, unless either the delayed or prompt circuit is tripped, in which case it bleeps intermittently even after RESET A goes high. After the alarm is set, it will bleep intermittently when the delayed circuit is tripped. Double check that all the alarm facilities are working, then remove the 100k resistors in parallel with R6, 24 and 28. Re-test the alarm again, but now all the timing periods should be the full length (approx 30 seconds delay for the delayed circuit and exit timer, and 18 minutes for the siren).

The case you use for this project will depend on how you intend to wire up the rest of the system; if you plan to run individual cables from each sensor switch back to the alarm controller, room is needed in the case for connector blocks, etc. Alternatively you can design the installation so that sensors are connected to one another, and only one or two wires need to go back to the controller. If the case is easy to open, you may want to include a case microswitch in the 24-hour loop. Position this switch so that its contacts open when the case is opened.

If you can find a three-way key-operated switch (I couldn't), this can be used as a more convenient alternative to SW1 and 2. Fig. 8.36 shows how to wire such a switch (the diode prevents the auxiliary sounder from operating while the alarm is disabled).

Installation

A useful source of advice over types and positioning of sensors is your local police crime prevention officer (CPO). I have seen some appalling 'professional' alarm installations, so don't be too put off if your CPO is dubious of amateur installations.

Many different sorts of sensors are available, from magnetically-operated door switches for fitting on door rims, through pressure mats, to sophisticated movement sensors, which can be either active ultrasonic or passive infra-red. Movement sensors require a source of power, and up to 100mA total can be drawn from the alarm's 12V power supply, using the connections at the top and bottom of the circuit connector blocks. Fig. 8.37 shows one possible arrangement using a variety of sensors and switches.

Decide on where to run the wires; they need to be kept out of the way, not just to avoid any intruders trying to cut them but also to prevent accidental damage. Run a loop of wire through any critical parts of the circuit which may be vulnerable to attack– for example, any wires a thief might attack before actually breaking in. At all times keep the delayed, prompt and 24 hour circuits separate, or the alarm will not work properly.

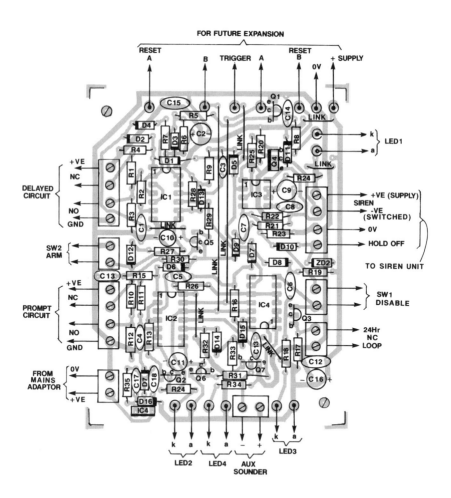

Fig. 8.34 Where the components fit on the PCB.

TABLE 8.2

Parts List: Alarm controller

Resistors (all ¼W 5%)

R1, 8, 10, 14, 18, 26, 27, 34	1k0
15, 16, 19, 20, 21, 22, 30, 31	R2, 3, 4, 5, 7, 9, 11, 12, 13, 10k
R6*, 28*	1M0
R17, 33	4k7
R23, 32	100k
R24*	10M
R25	220R, ½W
R29	3k3
R35	680

Capacitors

C1, 4, 8, 13, 16	10n ceramic 12V working
C2, 10	47µ 12V single ended low leakage or tantalum electrolytic
C3, 5, 6, 7, 12, 14, 15, 17	100n ceramic 12V working
C9	100µ 12V single ended low leakage or tantalum electrolytic
C11	2µ2 12V single ended low leakage or tantalum electrolytic
C18	100n ceramic, 30V working
C19	470µ 12V electrolytic

Semiconductors

IC1, 2	4069 (CMOS quad two-input NAND gate with Schmidt trigger inputs)
IC3	4082 (CMOS dual four-input AND gate)
IC4	555 (NE555 or equivalent)
IC5	7812 (12V 1A positive regulator)
Q1, 2, 3, 5	BC1168C, BC182L, BC183L, BC184L, 2N3704, 2N3705, 2N3706 or similar (small NPN)
Q6	BC212L, BC213L, BC214L, 2N3702, 2N3703 or similar (small PNP)
Q4, 7	BD135, BD139, MJE3001, 2SC1162 or similar (medium power NPN)
D1-10, 12-15	1N914 or 1N916 (signal diode)
D11, 16, 17	1N4001 (1A rectifier)
ZD2	BZY88C2V7 (2.7V zener, part of alarm protector circuit)
LED1-4	single LEDs to choice, with holders

Miscellaneous

SW1, 2	Single pole single throw key-switches (or one single pole three-way key-switch)
CB1, 2	Four-way PCB connector block
CB3, 4, 5, 6, 7, 8	Two-way PCB connector block
Siren	12V 250mA max (or up to 500mA with relay), minimum output 100dB at 1m
Secondary sounder	12V warning buzzer, low output, 100mA maximum current

Case to choice; case microswitch, single pole closed when case closed; unregulated mains adapter rated at 12V 1A minimum output; PCB; four-core burglar alarm cable; door switches, sensors, etc.

Fig. 8.35 The control unit housed in its box. The holes are for wires to the sensors, siren, sounder and power supply.

Modification for use as a car alarm

A simplified version can be used as a car alarm controller. The circuit of the car alarm is shown in Fig. 8.38 and the overlay diagram is Fig. 8.39. The same part numbering is used as with the main project, but the values of R6, 24 and 28 are changed. Other changes needed are:

* The 24 hour circuit is left out
* Much of the reset circuit and SW2 is dispensed with; the alarm is turned on and off by SW1
* The supplementary sounder and its circuitry is not needed, but the oscillator round IC1d and Q7 is modified to flash an LED on the dashboard to make the alarm's presence more obvious
* The delayed circuit's time is reduced and the prompt circuit is probably not needed, but can be retained if you wish

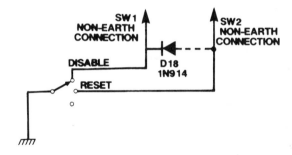

Fig. 8.36 Using a three-way key-switch instead of two two-way types.

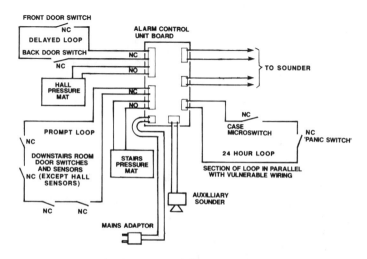

Fig. 8.37 One possible arrangement of the alarm.

* Any unused gates have their inputs connected to ground or the 12V
  supply

Installing the car alarm

Most cars have a courtesy light switch which can be wired to the alarm as in
Fig. 8.40. It may be worth fitting courtesy light switches to all the car doors,
if they aren't already attached. Ultrasound detectors and motion detectors
are also available for cars (house types are far too sensitive).

I suggest using a 12V relay, its coil in parallel with the siren, to make the
hazard warning lights flash when the alarm goes. Cars differ greatly in their
wiring, so you will have to look closely at the wiring diagram to decide how
to attach the relay contacts to the hazard warning flasher. Switch SW1 can
be mounted inside the car, well concealed, or on the outside as a key-switch.

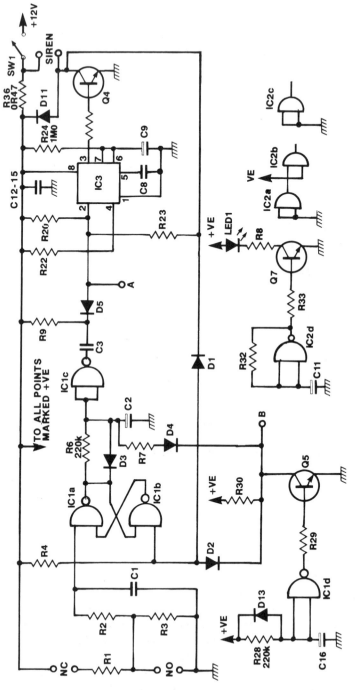

Fig. 8.38 The car alarm version.

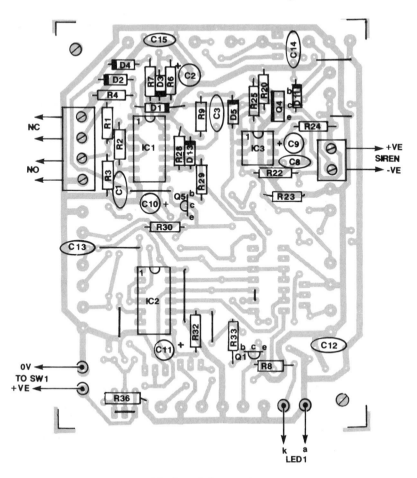

Fig. 8.39 The component positioning of the car alarm.

The values used for the timing give about 10 seconds to leave and enter the car; increase R6 and 28 for longer.

Finally, spare a thought for your neighbours. If any alarm goes off persistently, find out why as soon as possible.

PROJECT: Alarm protector

This is a medium project. Making the project itself is easy and involves some soldering. It can be used with the alarm project in this book or with many other alarms (see 'Is your alarm suitable' page 187). However, connecting it to an existing alarm will require considerable care.

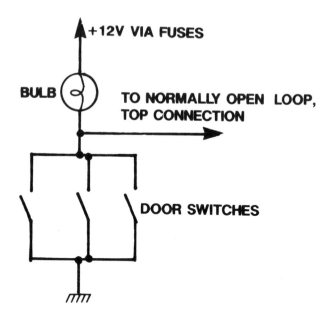

Fig. 8.40 Connecting the car alarm to the courtesy light.

This project improves the security of alarm systems which use only two wires to connect the siren unit to the control box. The problem with this type of system is that if a burglar cuts the cable to the siren, the alarm is completely disabled. The most interesting feature of the circuit is the way diodes and zener diodes are used to achieve functions which normally require more complex circuitry.

The circuit diagram of the project is shown in Fig. 8.41. Two transistors act as switches to connect the negative terminal of the siren to the 0V supply, completing the circuit and activating the sounder. These two transistors are in parallel with each other and also with the control unit's own switch in the control box, so that if only one of these switches closes, the siren is energised. (The control unit's switch may be a transistor like Q1 and Q2, or a relay contact.)

Transistor Q1 is used to defend the siren box. Current comes to the transistor's base via resistors R1 and R2. When the microswitch is open, the transistor turns on and sets off the siren; however, if the microswitch is closed, no current gets to the transistor's base so it remains off.

Transistor Q2 defends the cable. Current flows to the base of Q2 via R1 and zener diode ZD1, turning the transistor on. However, when the cable is intact, zener diode ZD2 is connected between point A and 0V, reducing the voltage at point A to 2.7V. No current should flow through ZD1, as there is

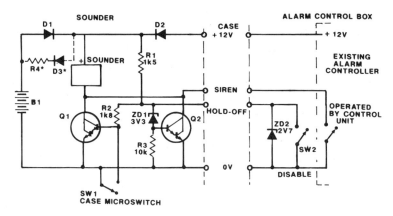

Fig. 8.41 The circuit of the alarm protector.

not sufficient voltage across it; in practice a small leakage current does flow and R3 is included to take this away from the base of Q2. If the siren comes on very faintly without the alarm being triggered, use a lower value for R3 (try 4k7 or 2k2). To make it possible to get inside the siren box without the alarm being set off, a switch SW2 is included inside the control box. This shorts point A to ground, so neither Q1 nor Q2 can turn on.

Darlington transistors are used for Q1 and Q2, which need a much lower base current than conventional transistors. As a result the standing current through the circuit is 3mA, but a mains power unit should be used for the main alarm otherwise the batteries will be drained in a week or so.

If a thief cuts the sounder cable, the supply from the control box is no longer available, so the alarm protector has its own battery. This can be non-rechargeable (a PP9), which will need checking and replacing periodically, or rechargeable. Diode D1 allows the battery to supply the circuit when there is no power from the control box. R4 and D2 should be included only if a rechargeable battery is used to trickle charge it.

If the cable is cut or the siren box opened, the siren will carry on sounding until the battery runs out. How long this takes depends on the size of the battery and how much current the siren takes; a typical siren will take around an hour to drain a PP9 battery.

Is your alarm suitable?

If you want to use it with an existing alarm, you will need to check some details of your system. The control unit must have a supply of 12V to 15V and the siren unit should have its positive terminal permanently connected to the positive supply, and the negative terminal of the siren must be the one which is switched to make it go off.

Unless you already know the full details of the alarm unit, you will need to open up the control box; if it is a sealed unit, abandon the project because you will break it trying to get inside. More usually, it will be a metal case with a hinged door held shut by a couple of screws. Check that you have disabled the alarm before opening up the case, so you do not annoy your neighbours. Inside the box, identify the mains transformer and check to make sure you can't accidentally touch live parts of the circuit. Find where the wire to the siren connects to the control circuitry; this will probably be via a connector strip mounted on the printed circuit board.

Look for a large electrolytic capacitor on the PCB; this is almost certainly the smoothing capacitor, and its negative end (identifiable from the markings on the case) will be connected to the 0V line on the circuit board. Connect the negative probe of your multimeter to its negative end, switch the meter to a range suitable for reading 12V DC and measure the voltage on the other end of the capacitor. It should be between 12V and 20V; any more or less, and the alarm protector cannot be used with this control unit.

Leaving the negative probe of the multimeter attached to the negative end of the capacitor, check the voltage of the connection to the siren. They should both be around the same voltage as the positive end of the capacitor (for possible exactly 12V, if a voltage regulator IC is used in the circuit). If they are both 0V, a modified circuit can probably be used. Disconnect the siren connections from the terminal block (if it immediately goes off, then an alarm defender is unnecessary because the circuit is more complex than you thought!) and then do something that would set the alarm off. Check the voltages on the terminals normally connected to the siren, noting which one stays at the positive supply voltage and which one goes to zero volts (if both were at 0V, note which goes to +12V or whatever).

Reset the alarm and reconnect the siren. Remove the meter and close the case; check that the alarm still functions normally; if not, reopen the case and look for any accidentally disconnected wires – single-cored solid wires are particularly vulnerable to breaking off.

Construction
The siren case may already be sufficiently large to accommodate the alarm defender and the battery; if not, buy a case large enough to contain both them and the siren itself. Use the opportunity to make the siren unit more obvious to a potential burglar. Plan how to mount the siren, PCB, battery and microswitch so that they don't trap rainwater. The mounting of the switch depends on the siren case; some will already have holes drilled for a case microswitch. In the prototype, it was mounted on the base of the case and arranged so that the lever was pushed in by one of the case screws.

Start assembling the PCB (Fig. 8.42) by mounting the PCB pins and the connector block, followed by resistors, zener diodes and finally the transistors. Finish by soldering wires to the pins for the battery and the case microswitch. Use your multimeter to determine which terminals of the microswitch wires should be connected to the circuit; use two terminals that are unconnected when the switch's lever is out but connected when the lever is pushed in.

TABLE 8.3

Parts List – Alarm protector

Resistors (all ¼W 5%)
R1   1k5
R2   1k8
R3   10k
R4   see text

Semiconductors
Q1, 2   TIP122 (TIP127 for positive switching version)
D1, 2, 3   IN4001
ZD1   BZY88C3V3 (3V3 500mW zener diode)
ZD2   BZY88C2V7 (2V7 500mW zener diode)

Miscellaneous
SW1   case microswitch, closed when case is closed
SW2   single pole single throw miniature switch
B1   battery – see text

PCB connectors (one 4-way, 3 optional 2-way, all 0.2 inch pin pitch; PCB; bell case for sounder (if existing case not usable); four-core burglar alarm cable (if not already fitted).

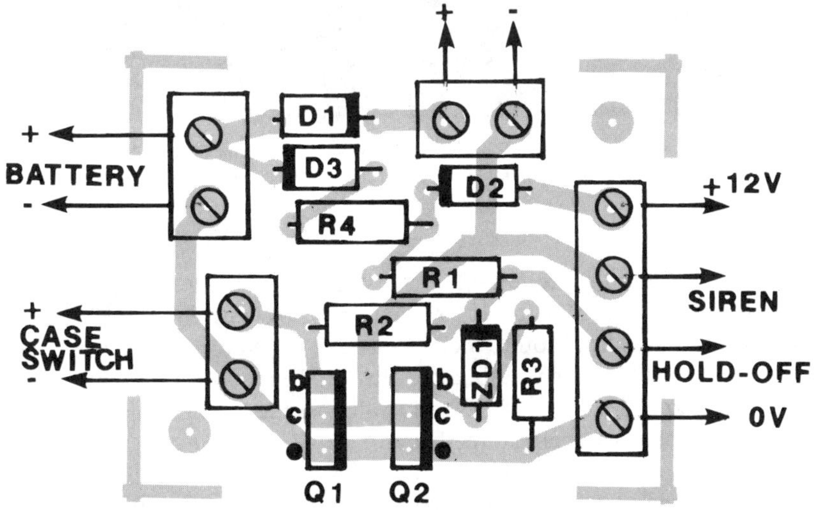

Fig. 8.42 Printed circuit board of the alarm protector.

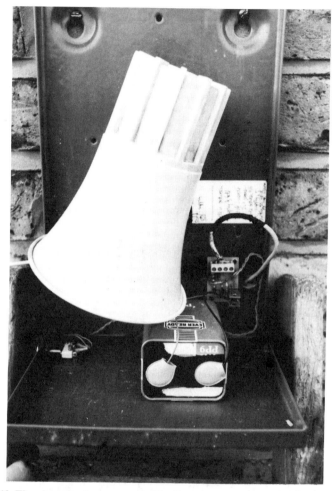

Fig. 8.43 The alarm protector mounted inside a siren box. The case microswitch here is at the bottom left inside the case.

Probably the most troublesome job will be replacing the double-core wire running from the siren to the control box with special four-core burglar alarm cable; if you're lucky, the cable might already be four-core type. Switch SW2 and zener diode ZD2 have to be mounted at the other end of the cable, inside the control box if at all possible, using a small bracket. The zener diode is soldered directly on to the switch; take care to get it the right way round. Cable colours are not shown on the circuit diagram, but the wires are numbered; if you use red for wire 1 at the control box end, you must use the same at the siren end.

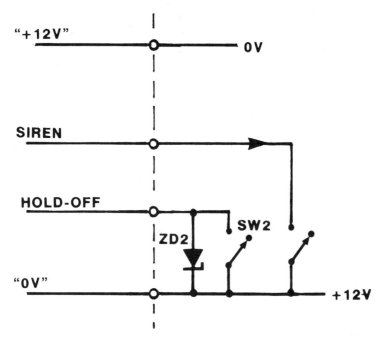

Fig. 8.44 Connections to the control box for the positive switching version.

Using rechargeable batteries

The alarm protector will not work with rechargeable batteries with a voltage of less than 7V, and preferably require a little bit more. Nickel-cadmium cells (nicads) have an EMF of 1.25V per cell, so six will give a voltage of 7.2V – rather close to the limit, so use seven to give 8.45V.

You must obtain cells with the appropriate capacity, in ampère-hours (AH) or milliamp-hours (mAH), and this depends on the siren current. Unless you have access to the specifications of the siren unit, you will have to set off the alarm with the multimeter inserted in the siren lead to measure the current the siren draws. It will probably be between 100mA and 1A, so switch the multimeter to the 1A DC range and connect it between the siren negative, disconnected, and the terminal at the connector block on the control unit where you've just removed the lead; warn your neighbours before setting off the alarm! If the siren current varies a lot, try to estimate the average.

The nicads should supply the siren for between 20 minutes to an hour. The capacity necessary for this is the siren current times the desired time (20 minutes, i.e. 1/3rd hour, and 1 hour). For example, if the siren draws 500mA, then choose a battery with a capacity of between $500 \times 1/3 =$

167mAH, and 500mAH. Nicads are available in many different shapes and capacities. There are AA style single cells – you will need seven of these and a special holder. Alternatively, there are rechargeable PP9 substitutes (check that the EMF is 8.45 not 7.2V) or ready-assembled stacks of 'button' cells.

Nicads are normally supplied discharged and must be charged before use. The alarm defender circuit is designed to trickle charge the batteries, and it would be much better to charge them before use in the circuit. This can be done by charging the cells from the 12V alarm supply via a current-limiting resistor. Appropriate values for the resistor are given in Table 8.4, and these give a charging time of about 12 hours.

Table 8.4 also gives values for resistor R4, based on battery capacity. If you use a battery with a different capacity (D) calculate the values for the 'quick charging' resistor Rq and R4 from:

$$Rq = 36/D \text{ and } R4 = 240/D$$

and take the nearest E12 resistor value.

Positive switching version

If the control box normally keeps both wires to the siren at 0V and takes one of them to around + 12V when set off, use the positive switching version of the circuit. This works in exactly the same way as the standard version, but with the whole circuit turned upside-down to operate with reversed voltages. The changes are:

1 Use TIP127 transistors for Q1 and Q2 (these are PNP transistors)
2 Reverse all the diodes from the directions shown in the overlay diagram
3 The connections to the battery must be reversed
4 Connections to the siren must also be reversed
5 The connections in the control box have to be changed: see Fig. 8.41.

TABLE 8.4

Charging resistor values			
Cell type	Capacity (ampère-hours)	Rq (ohms)	R4 (ohms)
AA	0.5	56*	390
PP9	1.2	36*	270
Button cells (8.4V)	170mAH	220	1k5

*Use a 1W resistor

CHAPTER 9

What to Do Next

What next? To answer that question, you need to decide if you want to keep electronics as a hobby or whether you would like to make it all or part of your career.

There are a growing number of courses at different levels, from GCSE to as high as you want to go. Colleges of all sorts offer electronics, and any course should prove a grounding for getting a job or getting on a course at a polytechnic or university.

Making it a hobby

You may still find it useful to do a course even if you just want electronics as an interesting and useful hobby. Another source of help and encouragement could be a club – local libraries keep information on clubs in the area. Unfortunately, there are not very many pure electronics clubs, but there are a great many radio clubs around, some of whose members will be building their own radio gear and other gadgets. Even if you're not at all interested in radio, radio rallies are a good place to buy odd electronic bits and pieces – though beware, the prices charged are not always bargains!

Whatever your future plans, start making some electronic projects now if you have not already done so. Besides those in this book, a number of magazines regularly publish projects. Foremost among these is, of course, *ETI*.

Building projects

Before commencing any project, check you can get all the components. At the time of writing, all the components in this book are widely available, but electronics moves fast, and older items can become obsolete.

Choose projects carefully, know your own limits and extend them gradually, rather than taking great leaps into the dark. Nothing is more discouraging than building an expensive project that you don't understand and can never get working. However, don't be discouraged if you build something and it doesn't work first time. Very few of my own projects work the first time but nearly all do eventually. Do learn from the problems you experience, finding out why it went wrong, and eventually you'll be able to get it going; every time you do this, your understanding will increase.

You will accumulate a lot of used items that might just come in handy 'one day' but, I urge some restraint. It is usually worth keeping only fairly new components with reasonably long leads, and ICs that you know to be working. However, aluminium and thin steel containers can be very useful even when exceedingly battered. Bare old PCBs are useless, but strip-board is often recyclable.

Component suppliers

The following is a list of the suppliers used for everything in this book, except the odd nut or washer which was bought from the local hardware store. Most components are available through a number of other suppliers and it would be worth looking at ads in the electronics magazines.

Maplin Electronic Supplies Ltd, PO Box 3, Rayleigh, Essex, SS6 8LR, Tel. 0702 554155: publishes a large and comprehensive catalogue (1992 edition £1.60 for 500+ pages) which provides invaluable information. Nearly everything used in this book was bought from Maplin, and they are happy to deal with orders from outside the UK.

Electromail, PO Box 33, Corby, Northants, NN17 9EL, Tel. 0536 204555, is the small orders section of RS Components. Although geared towards professional users, Electromail can be a good source, if sometimes a rather expensive one, for the hobbyist.

Cirkit Distribution Ltd, Park Lane, Broxbourne, Herts EN10 7NQ, Tel. 0992 444111: component supplier particularly strong on radio items.

Three smaller suppliers who may have components (particularly semi-conductors) not stocked by Maplin are: Electrovalue, 28 St. Judes Road, Englefield Green, Egham, Surrey TW20 0HB, Tel. 0784 33603; Cricklewood Electronics, 40 Cricklewood Broadway, London NW2 3ET, Tel. 081 450 0995; and Marco Trading, The Maltings, High Street, Wem, Shropshire SY4 5EN, Tel. 0939 32763.

Wilmslow Audio Ltd., Wellington Close, Parkgate Trading Estate, Knutsford, Cheshire WA16 8DX, Tel: 0565 650605: specialist in loud-speaker drive units and kits.

Riscomp Ltd, Poppy Road, Princes Risborough, Buckinghamshire HP17 9DB, Tel. 084 44 6326: specialise in security items, door switches, sensors, etc.

Printed circuit boards

A full list of prices of the printed circuit boards used in this book can be obtained by sending a stamped addressed envelope (envelope plus two IRCs from outside the UK) to: The ETI Book of Electronics PCBs, ASP

Readers Services Department, Argus House, Boundary Way, Hemel Hempstead, Herts. SP2 7ST. A list of any modifications to the projects will be provided if any problems are found, but please note that the Readers Services Department will not be able to answer any technical queries themselves.

Index

(Please note 'ff' after the page reference indicates an extended entry)